CELEBRATING
50 YEARS
Texas A&M University Press
publishing since 1974

T0405592

Power

Books made possible by
Sarah '84 and Mark '77 Philpy

JOE HOLLEY

Power

HOW THE ELECTRIC CO-OP MOVEMENT ENERGIZED THE LONE STAR STATE

Texas A&M University Press
College Station

First edition

Opinions expressed in this book do not necessarily reflect those of Texas Electric Cooperatives manangement or directors.

∞ This paper meets the requirements of ANSI/NISO Z39.48-1992 (Permanence of Paper).
Binding materials have been chosen for durability.

Library of Congress Cataloging-in-Publication Data

Names: Holley, Joe, 1946– author.
Title: Power: how the electric co-op movement energized the Lone Star State / Joe Holley.
Other titles: Texas experience (Texas A & M University. Press)
Description: First edition. | College Station: Texas A&M University Press, [2024] | Series: Texas experience | Includes bibliographical references and index.
Identifiers: LCCN 2023041874 (print) | LCCN 2023041875 (ebook) | ISBN 9781648431562 (cloth) | ISBN 9781648431579 (ebook)
Subjects: LCSH: Electric Reliability Council of Texas—Powers and duties. | Texas Electric Cooperatives—History. | Electric cooperatives—Texas—History. | Rural electrification—Economic aspects—Texas. | Rural electrification—Political aspects—Texas. | BISAC: HISTORY / United States / State & Local / Southwest (AZ, NM, OK, TX) | BUSINESS & ECONOMICS / Industries / Energy
Classification: LCC HD9688.U53 T44 2024 (print) | LCC HD9688.U53 (ebook)

| DDC 333.793/209764—dc23/eng/20231003

LC record available at https://lccn.loc.gov/2023041874

LC ebook record available at https://lccn.loc.gov/2023041875

Cover & book design by Laura Forward Long

To my Texas Electric Cooperative friends and former colleagues:

Mike Williams, Martin Bevins, Eric Craven, Karen Nejtek, Charles Lohrmann, Chris Burrows, Carol Moczygemba, Lara Richards, Lisa Braud, Carol Powell, and Gino Esponda.

CONTENTS

A gallery of images follows page 141.

PREFACE

We are Texans. We know unruly weather. We live and work in a state punished by periodic hurricanes near our Gulf Coast. Among their number is the greatest natural disaster in American history, the Great Storm of 1900 that swept across Galveston Island, leaving in its wake unbelievable devastation and at least six thousand deaths. We have seen tornadoes corkscrew out of gray-green clouds and with terrifying force claim lives and property. Flash floods scour neighborhoods, whole towns. Lingering drought and triple-digit temperatures wither crops, not to mention the human spirit. Occasionally we endure a bone-chilling norther, but the vicissitudes of winter are more the exception than the rule in this vast state.

Until February 13, 2021, that is. That's when Winter Storm Uri smashed its way into the record books, disrupting lives and livelihoods all across the state and laying claim to being one of the worst natural disasters in Texas history. More than 4.5 million electric utility customers lost power during the storm. Twice that number either had no running water or had to boil their water for days after the storm had passed. Uri was likely the most expensive natural disaster in Texas history, costing the state more than $295 billion in damages. More than two hundred people died.

Six months after Uri, the Hobby School of Public Affairs at the University of Houston conducted a survey of 1,500 Texans served by the Electric Reliability Council of Texas (ERCOT). The school wanted to know how Texans felt about their electricity providers' response to an unprecedented crisis.[1]

The answer? Not very good—unless their provider was one of the state's sixty-six electric distribution cooperatives (electric co-ops for short). "Overall, electric cooperatives significantly outperformed their rivals in the eyes of their customers during the winter storm of 2021," the Hobby School concluded.[2]

The survey found that co-ops performed their core functions better than the commercial utilities; their customers, who were member-owners, were more likely than customers of other utilities to believe that the co-ops had their best interests at heart. They also believed that co-ops were better able to respond to crises like the February storm than were deregulated electric utilities.[3]

As a former editor of *Texas Co-op Power* magazine, I was pretty sure I knew why. I was aware of how electric co-ops had originated more than eighty years earlier during a much longer and more pervasive crisis. Electric co-ops, unlike investor-owned utilities or municipal utilities, have been owned from their Depression-era beginnings by the people they serve. That sense of ownership and responsibility alone makes them responsive, during times of crisis as well as every day.

Kathi Calvert, general manager of Crockett-based Houston County Electric Cooperative, confirmed my suspicions. During an interview for a *Texas Co-op Power* article I wrote about the Hobby School survey, she reminded me that Houston County's East Texas consumer-members knew that the co-op folks were right there with them, experiencing the same misery and hardships they were experiencing. They would not have known that about large, anonymous utilities headquartered who knows where. They would not have run into their CEOs or their board members at church on Sunday mornings or at high school football games on Friday nights.[4]

In stark contrast, they saw co-op employees leaving their own dark and powerless homes and making their way to work with several inches of snow and ice covering the ground, temperature zero degrees. They saw bucket trucks on roads and streets and lineworkers in heavy coats clambering up ice-encrusted poles. "Collectively, we all survived," Calvert told me.[5]
She made sure that when anxious customers called in, they got their questions answered, even if the answer about such things as rotating outages might not have been what they wanted to hear. She had HR people, accounts-payable people, whoever might be available answering phones and keeping customers informed. She also made sure social media was providing the latest information. "It was a team effort, a community-based effort," Calvert said. "That's why co-ops are trusted."[6]

I found additional evidence from another part of the state. Joyce Buchanan, who had recently moved to McKinney—from Ontario, Canada, of all places—was quick to compliment her electricity provider, Grayson-Collin EC, as it coped with the storm. "As a recent transplant to Fannin County," she wrote, "I just wanted to say how impressed I have been with your updates and communication this week. They have been timely, informative and so helpful in letting us know what to expect from day to day, sometimes hour to hour." (I asked Buchanan whether she had any Canadian cold-coping advice for her new Texas neighbors. "Layer," she replied.) Brittany Brewer, a Fannin County EC member, echoed Buchanan. "We are lucky to have such a transparent power provider," she wrote in a Facebook message to her co-op.[7]

Cameron Smallwood, chief executive officer of United Cooperative Services, a Burleson-based distribution co-op serving North Texas, told Texas lawmakers a similar story during testimony before legislative committees shortly after the storm. United not only prepared members in advance for the likelihood of debilitating winter weather, Smallwood explained, but also used every means of communication available to keep its members informed. Communication is "part of our DNA," he said. "Our understanding is that customers from other utilities were watching our social media and information, because they were lacking information [from their providers]."[8]

"I'm one of your members," state representative Shelby Slawson told Smallwood. "We've heard a lot about the importance of communication with the public. I want to openly commend you and United Co-op for the way you handled that," the Stephenville Republican added.[9]

Coleman County Electric Co-op is a small West Texas co-op whose territory includes the aptly named town of Winters (home of the Blizzards). General manager Synda Smith said that she and her colleagues did their best to keep the lines of communication open even though the lines of power were shut off. "A good lesson learned," she said, "was that most people are more understanding if you keep the communication lines open and give them updates as to what is going on. Many understood that the situation was beyond our control and thanked us for taking care of them."[10]

Julie Parsley, chief executive officer of the pioneering Pedernales Electric Co-op, reported to her board of directors a few weeks after life had pretty much returned to normal. She recalled that co-op linemen and generation workers "were doing dangerous jobs in difficult conditions" during 165 consecutive hours of below-freezing temperatures. They were working sixteen-hour shifts in temperatures colder than Anchorage, Alaska. IT personnel who had lost power at home worked out of their cars, member-relations agents stayed in hotels close to PEC offices, and the co-op's Urgent Team was on the job 24-7, dealing with snow, ice, and mud even after the storm subsided. Systems and equipment occasionally failed, Parsley reported, "but the spirit and the resiliency of our employees surpassed that. . . . Our next step is to bring our systems up to the level of our employees, frankly."[11]

So why did electric cooperatives do so much better than their private utility cohorts? Those who conducted the survey—Kirk Watson, former dean of the Hobby School (as well as a former state senator and Austin mayor); senior director and researcher Renee Cross; and Rice University political scientist Mark Jones—suggested that the co-ops were more committed to their members' well-being than private utility companies were. Co-ops, in other words, prioritized their customers' interests, as they have from the beginning.[12]

When my dad was growing up on a Central Texas farm in the early years of the twentieth century, the life he lived, the life his parents and four siblings lived, would have been familiar to ancestors bound to the land centuries earlier in Wales, England, and Ireland. As Robert Caro so memorably conveys in "The Sad Irons," a chapter in the first volume of his LBJ biography, farm life in Texas and most everywhere else in America was hard and tedious. Since it was totally reliant on fickle Mother Nature, it was often hopeless.[13] "It was like stepping into another country," the writer Marquis Childs wrote in 1952, recalling rural Iowa when he was a child.[14]

It's little wonder that as a teenager my dad—and eventually all four of his siblings—fled the dawn-to-dusk drudgery. Hopping a freight before dawn one morning in the early 1920s, before the rest of the family awoke, he headed westward to San Angelo, where he moved in with his Aunt Ferel and Uncle Joe. His gregarious uncle was known to most of his San Angelo neighbors as Big Joe, an apt nickname for a fellow who stood six feet four and weighed three hundred pounds. With Prohibition still the law in the 1920s, Big Joe was a card shark, high-stakes gambler, and big-time bootlegger who ran a nightclub on the highway to Christoval called the Lone Wolf Inn. Most of his fellow Angelenos knew about his extralegal activities but didn't care, although a search of newspapers.com throughout the 1930s suggests that the authorities cared. The *San Angelo Standard-Times* reported frequently on his extralegal endeavors.

"I could set you up out here," Big Joe told his favorite nephew, "but Dell would never forgive me." (Dell was Joe's sister, my grandmother.)

Horace Holley didn't become a West Texas bootlegger and gambler; he didn't become a Central Texas farmer, either. Growing up bereft of electricity and running water on a blackland farm where the family raised cotton, corn, and watermelons, my dad knew from an early age there were less frustrating ways to make a living.

That yearning for something better meant that his three sons in years to come missed out on some of the good things that life on the farm had to offer: the sense of community with other farm families, learning Mother Nature's ways, a feeling of self-reliance, peace and quiet. My brothers and I grew up in working-class suburbia. (Our Hill County grandfather, by the way, didn't get power until the 1940s; he got running water a decade later. Today, Itasca-based Hilco Electric Cooperative provides power to the farms and rural homes near the Peoria community, where my dad grew up and where my grandfather lived for more than sixty years.) Even those who stayed on the land knew their lives could be easier, more productive, more fulfilling. They got a hint of how life could be every time they went to town, where the lights had come on in residences and commercial establishments decades earlier. They realized that electricity was a solution to the drudgery that was their lot, but it was a solution just beyond their reach. City people had it, but not their country cousins. Darkness settled in as soon as they passed the city limits sign.

Private power companies were not interested in extending wires out to sparsely populated areas beyond the city limits. There simply was no profit in stringing lines to scattered farms, ranches, and small communities—or so the companies maintained—which meant that farm and ranch wives continued cooking on woodstoves and their husbands trudged to the barn before sunrise and after sunset to milk in the dark. Their children did their homework (after chores) by the flickering light of a smoky kerosene lamp. Affordable electric power would be their salvation.

In *The Farmer Takes a Hand*, Childs describes the electric cooperative movement in this country as "an astonishingly swift social revolution." Paved roads, automobiles, and tractors made life easier for rural Americans, he noted, but electric power transformed life itself.[15]

The electric co-op story in Texas and throughout the nation is a story of neighborliness and community, of grit and determination, of persuasion and political savvy. It's the story of a grassroots movement that not only energized rural Texas but also "empowered" residents, reminding them of the strength they had when they banded together. Working together, they were Texans in the tradition of Stephen F. Austin's Old Three Hundred, of nineteenth-century German immigrants to the Hill Country, and of centuries-old Mexican

American farming communities relying on community acequias along the Rio Grande, as well as other Texans who have found strength in unity.

During my *Texas Co-op Power* days, some of the co-op pioneers were still around. The stories they could tell—and did tell—still empowered younger co-op members. Fortunately, a number of co-ops have compiled oral histories, some of which are included in this volume. As valuable as those origin accounts are, however, they are only part of the story. Electric co-ops have evolved over the decades, as Texas has evolved. Their future is arguably as intriguing as their past.

I should point out that I've been writing about Texas for many years, particularly small-town Texas. A number of stories I've written for *Texas Monthly*, *Texas Co-op Power*, the *San Antonio Express-News*, the *Texas Observer*, and other publications—most recently as the "Native Texan" columnist for the *Houston Chronicle*—have been adapted for inclusion in this history of the state's electric co-ops. Texas history in general and co-op history in particular are intertwined.

In my reporting and research for this book, I've also come to realize that the electric cooperative story in Texas is, ultimately, an American story. It's a nourishing source of pride in the past and an energizing inspiration, indeed a model, for the future.

For most Texans, your energy provider is just that—the private company or the municipal utility that keeps the lights on, that keeps the AC running when Texas temperatures hit the century mark, the faceless entity that sends you a bill every month. Texans who get their power from electrical cooperatives, if they're at all familiar with co-op history, are aware that their power provider is still, both in origin and intent, something akin to a movement.

Texas and Texans were an integral part of the national story—the national movement—from the beginning. A crusty US House speaker from Bonham named Sam Rayburn and his protégé, a young congressman from the isolated Texas Hill Country named Lyndon Baines Johnson, were key players in the long and arduous effort to turn the lights on in the countryside. The first house in the United States energized with power financed under the federal Rural Electrification Administration (REA) was near the small Central Texas town of Bartlett.

The residents of the farming town between Taylor and Temple had enjoyed electric service since 1905, but their neighbors who grew cotton and corn and raised livestock in the surrounding countryside stayed in the dark, stayed captive to rural drudgery. It wasn't until 1936 that Charles

Saage (pronounced "soggy") and his family had the honor of throwing the switch and electrifying the Saage farmhouse just outside town. The house is gone, but Bartlett Electric Cooperative continues to serve more than ten thousand members.

Today, Bartlett EC is one of seventy-five co-ops in Texas; they provide affordable electric power to nearly three million members throughout the state. The co-ops function in 241 of the state's 254 counties and are part of a venerable nationwide movement that serves more than forty million people in forty-seven states.

Bartlett is a good example of the evolving nature of the co-op movement and the expanding opportunities for electric co-ops across Texas and the nation. The small town with its red-brick main street and sturdy commercial buildings from the 1920s is still small, but it's also on the northern edge of the ever-spreading metropolis of Austin, one of the fastest-growing metropolitan areas in the nation. Bartlett Co-op still serves farmers and ranchers, but it also serves a growing number of customers who live in and around Bartlett and other small towns in the service area but who work in town—in Austin's huge high-tech community, for example. At Southwestern University in Georgetown. At Fort Cavazos near Killeen, the largest army installation in the country. As part of the growing Scott and White medical complex at Temple. At a growing number of high-tech enterprises spreading across yesterday's cotton and corn fields. Bartlett is also the energy provider for new businesses transforming Central Texas.

Co-op members working in these and other fast-growing cities within a hundred-mile radius of Bartlett, while enjoying a rural lifestyle, are unlikely to be milking cows before daylight or harvesting maize at sundown. Their energy needs are different from those of their traditional country cousins. This changing nature of rural and suburban Texas across the huge state offers challenges, indeed opportunities, to apply the co-op model to meet new energy needs. Those needs include high-speed internet service throughout the state, whether in the vast reaches of the Big Bend or the rapidly growing Metroplex, stretching these days from Waxahachie, south of Dallas–Fort Worth, to Sherman-Denison near the Oklahoma border, nearly a hundred miles away. The co-op model can provide distance education, telemedicine, and opportunities both to work from home and to revitalize small, economically struggling Texas towns.

Texas writer Bill Minutaglio and his wife, Holly, a UT-Austin administrator, live in Austin, but they spend quite a bit of time at their second home on the banks of the Llano River between Mason and Llano. One morning Bill was perusing his three-acre property when he happened to notice that a utility pole intruded on his view of the sparkling water riffling past huge, flat boulders. He called Central Texas EC, and a lineworker drove out later that day. Bill asked whether the pole could be relocated.

"We could do that," the lineworker said, standing with Bill and staring toward the river, "but, you know, there was a time when that pole and that line were things of beauty to people who live out here in the country."

Bill looked at the pole, looked back at the lineworker. "You know, you're absolutely right," he said. "Leave it right where it is."[16]

The late Al Lowman, a Texas writer and rare-book collector, would have known exactly what the lineworker was talking about. Lowman grew up on a coastal-plain cotton farm; he liked to say it was south of Violet on the road to Petronila (both near Corpus Christi). "My earliest memories," he recalled, "are of Aladdin lanterns, battery-powered radios, treadle sewing machines and mud roads."[17]

The Lowman family finally got electricity two weeks before Christmas 1938. That's when the newly formed Nueces Electric Cooperative, with the help of a young congressman named Lyndon Johnson, got a $400,000 loan from the newly established Rural Electrification Administration (REA). For the first time ever, the Lowman family Christmas tree was strung with multicolored lights. Four-year-old Al was standing in the front yard with his dad when co-op lineworkers finished connecting the house to the newly installed power line. Innumerable rural Texans experienced a similar sense of hope and elation.[18]

"Look back at the house," the elder Lowman said. The youngster turned in time to see the lights flash on. He raced into the house to stand before the Christmas tree, strings of light gleaming from the lowest branches to the star-topped tip.

"It was a moment of sheer magic," Lowman told an annual meeting of the Texas Folklore Society nearly a half century later. Nothing else in his lifetime matched that moment, he said.

For the Lowman family and their South Texas neighbors, for rural Texans across the vast state, the electric co-ops were formed to meet a need. The co-ops have evolved over the decades to meet evolving needs. They haven't always been perfect: scandals have erupted now and then, most spectacularly within the nation's largest, Pedernales Electric Co-op. Mismanagement within individual co-ops has occasionally handicapped the vital work they do. And the lack of effort until fairly recently to enlist Texans of color in the co-op story, as well as women, has been a glaring oversight.

Despite the occasional shortcomings, the electric cooperatives are heirs to a proud tradition of public service. For the most part, they have stayed true to their foundational aim: empowering people to improve their quality of life. This is their story.

Power

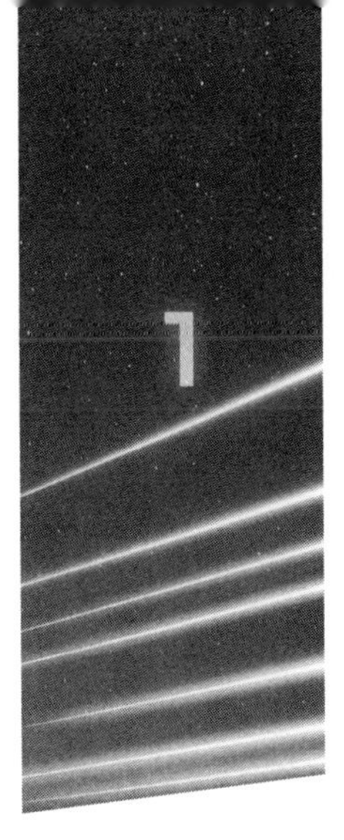

"CAUGHT AND . . . CHAINED"

On July 4, 1885, the city of Austin staged an enormous Independence Day celebration to coincide with the placement of the cornerstone for a new hotel at the corner of Pecan Street and Brazos. It was called the Driskill in honor of its builder, a wealthy cattleman and longtime Austinite named Colonel Jesse Driskill. Busts of the builder and his sons would adorn the facade.

The colonel had announced that his hotel in the heart of downtown, six blocks south of the capitol, would be "the most sophisticated hotel west of St. Louis" and would cost him $400,000 to build. Four stories high, the hostelry would feature hydraulic elevators and flush toilets. Guests could lounge on a large couch in their rooms or relax in a red rocker. Retiring for the evening, they could sleep in a four-poster bed. If they needed anything, an electric bell would summon a porter. Illuminating each of the hotel's sixty rooms on the top two floors would be elaborate electric-powered chandeliers.

The *Austin Daily Statesman* reported that between four thousand and six thousand Austinites gathered for the Driskill ceremony beneath a string of incandescent lights on Pecan Street (now Sixth Street). They listened to a brass band, a string band, and speeches by Driskill, Austin mayor John W. Robertson, and other dignitaries. An anonymous reporter noted that the ceremony was illuminated not only by electric lights but also by "the beauty of Austin," including the women of the city. "As long as the reporter started in by calling attention to the young ladies," he wrote, "he cannot close this report without saying point blank that if all the young and handsome ladies last evening reside in Austin, the city well sustains its boasted reputation of having more pretty ladies and ugly men than any city on the American continent."[1]

Those men and women, attractive and otherwise, listened to a keynote address by a prominent Austin lawyer and land agent named Edward W. Shands. An Indiana native who had moved to Austin in the 1850s, Shands invited his audience to imagine Austinites at the same downtown intersection 115 years into the hazy, distant future, July 4, 2000. The occasion, he said, would be the razing of the by-then venerable old hotel. Shands predicted that a quarter of a million people would be living in their fair city in the year 2000. He imagined them looking back on their forebears, the Austinites of 1885, and shaking their heads at how those earlier Austinites would respond to their city at the dawn of the twenty-first century. "What would be their amazement if they could only witness the wonders which have been accomplished since electricity has been caught, and, in a measure, chained."[2]

"Electrical airships," Shands envisioned, would be ferrying passengers "from Austin to San Francisco, from thence to China and Japan, returning over Asia and Europe and across the Atlantic to home again . . . making a pleasure trip around the globe in a few days." Electricity would be used "to send shock waves through people, causing them to live longer and end all diseases," Shands imagined. Americans of the future would be safe not only from disease thanks to electricity but also from foreign invaders. Shands predicted that "entire armies and navies" that dared attack America would be "instantaneously destroyed with one electrical bolt!" Every day, sixty thousand copies of the *Austin Daily Statesman* would be delivered through pneumatic tubes to every building in Austin. The Postal Service would be delivering the newspaper to home subscribers; electrical airships would drop the papers "upon the breakfast tables of subscribers, residing hundreds of miles distant from Austin."[3]

Austin's ambitious hotelier was putting the miraculous power of electricity to work less than a decade after an eccentric genius based in New Jersey had made electricity possible for use in homes, industry, and transportation. His name, of course, was Thomas Alva Edison.

Born in 1847 in Ohio and raised in Michigan, Edison had been conducting experiments in his family's basement ever since his mother had given him a book about chemistry. She worried that her gifted son would burn down the house, so at thirteen, he emerged from the basement and got a job as a "train boy," selling snacks and newspapers on the line between Port Huron and Detroit. He also built a small chemistry lab in a baggage car, allowing him to indulge his curiosity about the wonders of science between stops, and he printed his own newspaper on the train. The newspaper was a moneymaker for the young man.[4]

One day he happened to spot a stationmaster's three-year-old son playing on the tracks and pulled him to safety almost in the shadow of a fast-approaching train. As a reward, the boy's father taught young Al the wonders of Morse code and showed him how to operate telegraph machines. Not long afterward, Al's baggage-car lab caught fire, and the conductor kicked him off the train for good.[5]

Finding occasional jobs with Western Union and other companies, the young man knocked around Kentucky, Ohio, Indiana, and Tennessee for a few years. In 1869, he patented his first invention, an electric vote recorder that tallied votes instantly. Although he found no buyers for the machine, it was the first of the 1,093 inventions, processes, and systems that Edison would patent before his death in 1931.[6]

More inventions quickly followed, including a stock-ticker device and a telegraph machine capable of sending four telegrams simultaneously. Western Union got interested in what else the young man might come up with, so they paid him to experiment. Using that money, he bought a two-story building on thirty acres near Newark, New Jersey—his laboratory and machine shop would be called Menlo Park—and set to work tinkering and inventing and also perfecting what others already had invented (including incandescent lights). Menlo Park also included a three-story house for his wife, Mary, and their three children, two of whom he nicknamed Dot and Dash.[7]

Young Edison wasn't the first tinkerer-inventor obsessed with artificial light. The first was arguably Alessandro Volta, the Italian physicist and chemist who invented the battery in 1800.[8] Nor did Edison invent the electric light. The inventor of the first practical electric light was Humphry Davy, a British chemist from Cornwall who in 1802 developed the Davy lamp, an early version of the arc lamp. Davy discovered that a current passing through two carbon rods set side by side would produce an arc of white-hot light across their ends. In years to come, other inventors commercialized his invention, and in the 1870s it became the first electric light to illuminate town squares, hotel lobbies, and shop windows.[9]

Paris hosted the first International Exposition of Electricity in 1881, but the City of Light—named for its fifty-six thousand gas lamps—had little electric illumination beyond the fair. When European cities did adopt electric lights, they brought in American firms to do the wiring. Across America, inventors and would-be inventors were coming up with ideas for improving the nation's electrical system. Edison jokingly suggested that perhaps the British just didn't eat enough pie, although a more supportive patent system was a more likely reason.[10]

Suddenly, America glowed—in the towns and cities, that is. The bright lights of theater marquees drew audiences to evening shows. Electric lighting allowed surgeons to see more clearly, and lights attached to microscopes confirmed germ theory. Clothing and jewelry sparkled with tiny incandescent bulbs, although few matched ultrawealthy socialite Alice Vanderbilt. She was the talk of the 1883 social season when she arrived at a costume ball as the "Spirit of Electricity," her golden gown studded with diamonds and tiny bulbs lit by hidden batteries.[11]

It took a while for home lighting to catch up. The overwhelming brightness and flickering made the arc lamp less than ideal. Most homes relied on indoor gas lamps to defeat the darkness, but they produced fumes that permeated the house, and they covered everything in soot. A regular chore for many children was scouring the soot-covered lamps.

The incandescent light bulb lit by an electrified glowing filament was the solution in theory, but finding the right filament proved tricky. Most burned out too fast or didn't glow brightly enough (or both). Edison and the team of engineers and scientists he organized at Menlo Park figured out how to make a bulb burn longer and more reliably and at a level of brightness that wasn't hard on the eyes. Their biggest challenge was finding a filament that would burn slowly and steadily. Experimenting with six thousand plant fibers, they finally settled on carbonized cotton, which burned for fourteen and a half hours in the fall of 1879. A year later they got carbonized bamboo to burn for a thousand hours.[12]

As Michael Lind states in *Land of Promise*, "Edison's most important contribution was to create an entire system of electric lighting, from the generator through the wires to the incandescent bulb in the electric lamp. Edison announced that his goal was to make electricity 'subdivided so that it could be brought into private homes.'"[13]

At 3:00 p.m. on September 4, 1882, Edison activated a switch at his Pearl Street steam-electric station in lower Manhattan, and the age of electricity began. Pearl Street was the first electric power plant in the United States. (Another plant Edison had designed had gone on line in London some months earlier.) Pearl Street's initial customers included the financial offices of J. P. Morgan (who would arrange in years to come the merger that became General Electric), as well as the *New York Times*. The newspaper published a glowing account, so to speak, of the light emanating from fifty-two incandescent light bulbs in its newsroom: "The light was soft, mellow, and grateful [*sic*] to the eye, and it seemed almost like writing by daylight to have a light without a particle of flicker and with scarcely any heat to make the head ache."[14]

Pearl Street started with 119 customers. That number had grown to 500 within a year. "The revolution Edison had wrought was so unobtrusive and at the same time so world changing," biographer Edmund Morris wrote, "that few, if any, of the people who experienced it realized what had happened: an end to the counterbalance of night and day that had obtained for all of human history, mocking the attempts of torchbearers and lamplighters and gas companies to alter it with their puny waves of flame."[15]

Slightly more than a dozen years beyond Austin's Driskill Hotel celebration, airships were closer to becoming reality than E. W. Shands might have imagined. Although medical shock waves and military electric bolts were still figments of the fervid Shands imagination, electric motors were powering factories, trollies, and elevators (making buildings that scraped the sky possible). On May 23, 1898, *Daily Statesman* readers could only marvel at a house in Buffalo, New York, described in the newspaper's headline as the "Palace of a Magician," thanks to electricity.[16]

The so-called "electrical palace" belonged to one Frank C. Perkins, an electrical engineer who had filled his residence "from cellar to garret with electrical wonders." Perkins was a socialist city council member who never spent more than fifty dollars on any of his campaigns. Born in a small town in upstate New York, he studied electrical engineering at Cornell University and in Germany before establishing a consulting business in Buffalo. He conducted night school classes to teach streetcar streetcar operators the fundamentals of electricity and also published articles in electrical journals in the United States and Europe. As a socialist, he advocated municipal ownership of electrical power plants to light the streets and asphalt plants to pave the streets.[17]

Perkins also invented the first electric incubator to be patented. The Perkins residence was the first in Buffalo to be wired for electricity; Perkins, of course, did all the wiring. As soon as visitors to the residence set foot on the veranda, they activated incandescent lights on the veranda and in the vestibule. When the front door closed, the veranda lights went out and lights on the first-floor landing automatically came on. An oak panel on the landing contained switches behind a secret door that turned on lights upstairs in the hallway, the bathroom, or the attic. Callers at the front door could speak to a resident in the sewing room, the nursery, or elsewhere, who could admit the visitor by pressing a button. The front door opened automatically. The house was protected by an electric burglar alarm. If any door or window was forced open, the electric lights on the veranda and in the vestibule immediately flashed on, and an alarm bell rang in the kitchen.[18]

The temperature of the whole house was electrically maintained, the *Daily Statesman* reporter noted, "and after once being set in the fall, the heat is evenly distributed throughout the year." The kitchen was equipped with electric knife sharpeners, electric coffee grinders, and electric chafing dishes. "The lighting effects in this house are novel and interesting," the reporter continued. "No chandeliers are to be found in reception hall, parlor or dining room. The electric lighting is accomplished by electric ground glass globes in the side walls and ceilings, back of panels of chipped and opal glass shedding a soft, subdued light about the rooms with none of the fixtures of the past."[19]

The bathroom featured an electric shaving cup providing hot water. The foot-controlled sewing machine in the sewing room was, of course, electric, as were the irons in the laundry room. The den was equipped with an electric cigar lighter. "The designer of all these wonders is an electrical engineer," the *Statesman* reported. "As may be supposed, he makes his electrically fitted house the hobby of his life. It is safe to say that there are very few residences that equal it for labor saving and comfort giving devices."[20]

Nothing in Texas equaled the Buffalo house, but the state was quickly finding uses for electricity. Power arrived in the state in early 1882, when electric lights were installed at the Galveston Pavilion. Originally known as Saengerfest Hall and owned by the Galveston City Railroad Company, the structure at Twenty-First Street and Avenue Q was the city's first to be built specifically as a resort.[21]

The state's largest city at the time debated the power issue at a regular meeting of the city council on February 6, 1882. Galveston mayor W. R. Baker reported that to properly light the city with gas would require 1,700 lamps at an average cost of $4.90 a month for each lamp, or $8,330 in the aggregate, "which is nearly double the whole amount now collected of taxes for all purposes except the payment of interest; therefore, the purpose of a general lighting of the city, given all portions of it an equitable division of gas light would be too expensive and beyond reach."

Mayor Baker reported that to satisfactorily light the city with electricity would cost $15,435, and the cost of maintaining, operating, and furnishing the light would be $30 to $50 per day, or $7,435 a year, "from which can be deducted at least $500 for moonlit nights when the light need not be used and a portion of the daily expense can be saved."

The mayor proposed appointing a three-person committee, including himself, to investigate, although he noted that if the committee recommended electricity, the city didn't have the money to light the streets, except at the expense of other departments. He proposed a subscription to construct a power plant that would be donated to the city upon completion. He offered to personally contribute $1,000 toward that end.

The lights flared in Galveston and shortly afterward in Dallas, Houston, Fort Worth, and other towns and cities across the state. Hundreds of small electric companies were created, and generators were built especially to power ice plants, trolley systems, and cotton gins. These specialized power generators were often extended to surrounding homes and businesses to be used only at night. For example, power lines from trolleys were simply connected to homes from lines running along the street.

Back in 1881, the thought of electricity for Dallas was ridiculed as both dangerous and worthless, especially with the early failures to light London, New York, and New Orleans. Nevertheless, Alex Sanger, Julius Schneider, and W. C. Connor proposed bringing electric lights to the city when they incorporated the Dallas Electric Lighting Company in 1882. After receiving their charter from the state of Texas, they successfully petitioned the city of Dallas for permission to construct and operate an electrical system and quickly built the first power-generating system in Dallas in an abandoned wood-frame church on Carondelet Street (now Ross Avenue).[22]

Three locations hooked up, burning a total of twelve light bulbs. The 1883 City Directory described the illumination provided by the arc lights as pale and ghostly, with weird rays. Private funds were used to build the plant only four years after Edison announced the invention of the incandescent lamp and one year after he installed the dynamos to begin operation of New York City's Pearl Street Station.[23]

The *Dallas Daily Herald* welcomed the arrival of electricity. "Wherever there is an electric light, there is a bright and cheerful spot," the newspaper editorialized, "and the dark shadow beyond the light is anything but pleasant."

Electrical service began in 1883, lighting the streets and electrifying the streetcars. The most well-lit spot, according to Dallas Power & Light, was Mayer's Garden, one of the larger, better-known saloons on Elm Street in downtown Dallas. Service spread to homes and businesses as various electric companies built larger power stations.[24]

The Dallas Electric Company, chartered in 1890, constructed a massive brick power station on Griffin Street to the side of the Missouri, Kansas & Texas Railroad tracks. By 1899, the station, originally one story with a basement, was expanded with a brick and steel boiler house. At the time, it was the largest power plant outside New Orleans. In 1902, Dallas Electric Light & Power acquired the plant and property. In 1906, it built a new plant on the site to meet the growing electrical needs of the city.[25]

In 1900, the Dallas Street Railway Company substituted electric power for mules. In 1904, interurban electric service was inaugurated between Dallas and Fort Worth, but in April 1908, when the Trinity River reached a height of 37.8 feet, the power plants were underwater. Weeks went by, and by the time the flood was over, the river had crested at 51.3 feet in May. The city was dark, without water, other than from artesian wells, and with no fire protection.

In his acclaimed memoir, *This Stubborn Soil*, William A. Owens recalled leaving the succession of tenant farms in northeast Texas where he grew up in the early years of the twentieth century. At fifteen, he visited Dallas for the first time and got a job at the massive Sears Roebuck mail-order house on South Lamar. Young Owens had seen Paris (Texas, that is), but he had never seen anything like Big D. Dallas had lights and electric streetcars:

> "Look at the lights," Maggie said. "You could go a long ways and not see anything as pretty as Elm Street at night." I shook my head, too taken by what I was seeing to talk about it.
>
> Far ahead I saw the word majestic and above it a strange and beautiful bird in red and green and yellow with crane's legs that stepped

> up and down, up and down, not going anywhere but, staying where it was, its red and green topknot high against the sky, making me forget everything but the knowing that I was in Dallas and this was a part of Dallas. Without a word I went to the back of the car and watched till we passed the curve that separated the light and dark parts of Elm Street.[26]

Waco, ninety miles south of Dallas, was one of the state's leading cities in the late 1800s, but city fathers were apprehensive about electricity. On January 1, 1885, the city council decided to table the decision of converting to electric power. Writers from a Waco newspaper called *The Day* lamented, "Alas! For the prospects of electric lightening [*sic*] in Waco. They are dim." Nine months later, on September 18, 1885, the council voted 5 to 4 in favor of installing electric streetlights. "Fiat Lux!" the *Waco Examiner* proclaimed. The city was considering a "contract for a plant that will spread light over 15,000 people."[27]

Four days later, the council was still mulling its decision. Finally, an alderman named Alexander had had enough. "We have been gassing on the light subject long enough, now let us adapt electric light and approve the contract." His colleagues were persuaded. They approved a contract with an Indianapolis electric company, and work began. On March 1, 1886, Waco officials turned on their electric streetlamps for the first time. By 1892, three electric companies in Waco were providing power to local customers. Electric enlightenment arrived on the campus of Baylor University, south of downtown, in 1902.[28]

El Paso had been tinkering with electricity since the 1880s, as well. In 1883, local entrepreneurs experimented with batteries as a source of electrical current, but they couldn't make a go of it commercially. Three years later, the city erected a small direct generating plant that managed to provide current for a few arc streetlights and stores. By 1890, a prominent El Paso pioneer named Zach T. White had taken charge of the electrical energy business and had new generating equipment installed at Third and Chihuahua Streets. The use of incandescent lamps began in El Paso stores and homes.[29]

The state of Texas issued a charter to the International Light and Power Company in 1889 for the construction and operation of a generating plant to supply power not only to the city of El Paso and neighboring communities on the northern side of the Rio Grande, but also to Juárez.[30] The El Paso

Street Railway Company, chartered in 1882, became the El Paso Electric Railway Company on August 30, 1901. The mule-drawn streetcar system serving El Paso and Juárez switched to electricity. "The mule-car system was doubtlessly considered modern when established," El Paso Electric history points out. "However, as the community grew, it aspired to more modern conveniences. From the East came reports of electric lighting and streetcars that got their power from overhead electric trolley wires. Some El Pasoans believed that mule cars were too slow for an up-and-coming city."[31]

Mandy the mule, one of the last mules to pull street cars, may or may not have agreed. She was honored during an international ceremony marking the transition. Instead of pulling, Mandy got to ride on a flatcar over the new electrified streetcar route between El Paso and Juárez. From that day forward, electric streetcar routes branched out across the city from downtown to Fort Bliss and elsewhere.[32]

El Paso Electric Railway Company purchased the International Light and Power Company in 1905 and the electric portion of El Paso Gas and Electric Company in 1914. As the electric systems grew, alternating current replaced direct current, except for streetcars and some elevators.[33]

Life for power company employees in those days could be adventurous, as T. Elmer Hearn recalled in a 1951 interview with the *El Paso Times*. Repairing electrical facilities in 1910, he and his coworkers suddenly realized that bullets from across the river were splattering all around them as combatants during the Madero revolution fought it out. "Some of the men stood on top of the Santa Fe plant to watch the excitement, but the bullets began whistling by and they all ducked to safety," Hearn told the newspaper. "One man was killed as he stood on a porch across from the plant on Santa Fe Street," he added. "We were called to Juárez to repair lines, and every now and then some ardent revolutionist would start firing his rifle in the air. We made pretty good targets on those poles."[34]

By the late 1890s, residents of smaller towns like Hearne in Central Texas and Sherman in North Texas could flip a switch and enjoy the benefits of electricity. The *Brenham Weekly Banner* reported on November 11, 1897, that the small town of Terrell, east of Dallas, "has contracted for a fine system of electric lights." Weatherford, population a little over 2,000 in the 1880s, organized the Weatherford Water, Light & Ice Company in 1887. Giddings got power in 1905. Twenty years later, a privately owned plant was serving 148 customers.[35]

Power came to Abilene in 1891. Two years later, the *Abilene Reporter* noted distant rumors of "an important change in the method of municipal transport." Stages or carriages, the newspaper observed, might soon move London's populace about by means of electric power. "Storage batteries are to be used. No one . . . will deny that the perfection of the storage battery will make this possible."[36]

Entrepreneurs and shysters were sensing opportunity. On February 23, 1878, the *Galveston Daily News* advertised "reliable help for weak and nervous sufferers. Chronic, painful and prostrating diseases cured without medicine. Pulver Machine's Electric Belts the grand desideratus. Avoid imitations."[37]

On February 24, 1884, the *Fort Worth Daily Gazette* reported on the "Electric Lighter," a device that not only produced light but also functioned as a burglar alarm. "Our burglar alarm is so constructed that the intruder is immediately confronted with light and a bell alarm at the same instant. Reliable agents wanted all over the country."[38]

A full half century after Galveston became one of the first Texas cities to wire for electricity—through the Spanish-American War, the dawn of a new century, the birth of flight, the Great War, a worldwide pandemic, the Roaring '20s, and the subsequent devastating collapse of the economy—the countryside was still dark at night. In the early 1920s, Charles Lindbergh and other daring US Postal Service pilots could guide themselves across the continent at night by following clusters of light that marked Dallas–Fort Worth, Houston, San Antonio, El Paso, and Austin. Beyond the sporadic glow were broad swaths of ink-black darkness.

Although the pilots couldn't see it, farm and ranch families down below were living in the dark just as they always had. Dairymen were waking up before sunrise and trudging to their barns before dawn. Wives were firing up balky woodstoves and pumping well water by hand to begin preparing meals for the day. Children were still doing homework in the dim light of smoky, smelly kerosene lamps. Radio was practically unknown. So were washing machines, water heaters, refrigerators and freezers, vacuum cleaners, bathrooms, irons, electric pumps, and, of course, electric lights.

Robert A. Caro, an acclaimed biographer who spent decades trying to plumb the depths of LBJ, has written what is arguably the most profound description of the hardship and drudgery of daily life in the isolated Texas Hill Country before electricity. In "The Sad Irons," a chapter in volume 1, *The Path to Power*, Caro notes that "sad iron" was the name farm wives gave to a six- or seven-pound wedge of rough iron with a detachable handle, heated on a woodstove every few minutes and used to press out the wrinkles in their family's newly washed clothes. "Washing was hard work, but ironing was the worst," a farm wife told Caro. "Nothing could ever be as hard as ironing."[39]

Life was hard for wife and husband. Because there was no electricity, a Hill Country farmer had to milk and water his cows by hand. He had to grind kernels of corn for his mules and horses, a tedious chore that required stuffing corncobs one by one into a corn sheller and cranking it for hours. He had to unload cotton seed by hand, saw wood by hand, feed his livestock by hand, swing an ax by hand, and finish his chores after sunset just as he had begun the day before sunrise—"stumbling around the barn milking the cows in the dark, as farmers had done centuries before."[40]

All because there was no electricity.

"But the hardness of the farmer's life paled beside the hardness of his wife's," Caro writes. She had to carry heavy buckets of water from a stream or a well. She had to haul wood because Hill Country stoves were woodstoves. Maybe she had children to help her, but more than likely she was the one who cooked on that stove, who had to get the fire going and keep it going. She had to start every meal from scratch because there were no electric refrigerators to keep ingredients. She had to cook three meals a day, not just for her family but at certain times of the year for a twenty- or thirty-person harvesting crew. In the summer she had to can fruits and vegetables the very day they ripened, a sweaty, backbreaking job that required constant attention over a hot stove in a stifling kitchen. And then there was washday once a week, another backbreaking chore that required three washtubs of boiling water, water that had been carried to the house from the well or stream. As Caro noted, a Hill Country farm wife had to do her chores day after day, year after year, even if she was ill. Oh, and someone had to look after the children. "In the Hill Country," Caro writes, "the one almost universal characteristic of the women was that they were worn out before their time, that they were old beyond their years, old at forty, old at thirty-five, bent and stooped and tired."[41]

Because there was no electricity.

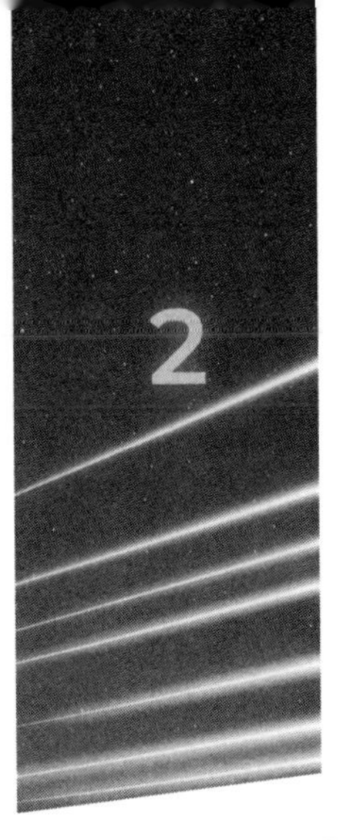

"BY THE WAY, HE ALWAYS PAID FOR HIS CHEESE AND CRACKERS"

In 1906, a twenty-four-year-old North Texan decided to run for the Texas legislature. To be a lawmaker (and a lawyer) was an unlikely dream for a young man who had grown up poor alongside ten siblings, his sturdy neck red from bending over long rows of cotton day after day under a punishing summer sun. On a forty-acre farm in Fannin County, near a hamlet called Windom, "I plowed and hoed from sun till sun," he often recalled in later years.[1]

Samuel Taliaferro Rayburn wanted something more—for himself, for his family, for his neighbors, who, like the Rayburns, also struggled to make a living on the fickle blackland soil. He vowed to make something of himself and had decided on the law and politics by the age of eight or nine. He would be one of those heroic figures he encountered in political biographies he devoured in the flickering light of a kerosene lamp, one of those politicians he read about in newspapers and liked to talk about with grown-ups when he got to ride into town with his father on Saturdays. Chopping wood or picking cotton, he imagined himself making speeches on a courthouse square, debating issues of the day in Congress.[2]

The earnest young farm boy paid his way through school at tiny East Texas Normal College (now Texas A&M University–Commerce) by sweeping out a nearby elementary school. He also held down a second job. As the college's bell ringer, he had to stop what he was doing every forty-five minutes and scurry up the bell tower to signal the end of a class period. Graduating in two years, he took a job as a teacher in a one-room school in a rural community near Bonham called Dial. Two years later, the ambitious young man wangled a better-paying job at the three-teacher school in a Fannin County settlement called Lannius.[3]

In 1905, the Texas legislature adopted the so-called Terrell Election Law, which meant that candidates for public office no longer would be chosen by powerful, often corrupt party bosses meeting in proverbial smoke-filled rooms. Instead, the people would choose, voting in direct primary elections.[4]

The aspiring young lawyer-politician teaching kids in a rural school saw his chance. Party-boss connections he didn't have, but that wouldn't matter. He could run as a candidate for the Texas House of Representatives in the next Democratic primary. He just might win the nomination, tantamount to election in the overwhelmingly Democratic state. As a lawmaker, he'd be making five dollars a day for the first sixty days of each biennial session and two dollars for each day of the session thereafter. For the young teacher, a legislative salary constituted a raise, maybe even enough for him to enroll in law school at the University of Texas.

In nearby Honey Grove, another Sam—Sam H. Gardner—had made a similar decision. Ten years older than Rayburn and better known around the county, Gardner declared his candidacy. With a long, hot summer coming on, young Rayburn set in to campaigning. Riding from farm to farm astride a little brown cow pony, he shook hands and talked issues with anybody who'd take the time to lean on his hoe and listen to the young man's views on Prohibition or boll weevil infestations or the declining price of cotton.[5]

As biographers D. B. Hardeman and Don Bacon tell the story, Rayburn had to spend time trying to convince farmers in their fields or folks downtown on a Saturday that he was old enough to run for public office. Hoping to appear more mature, the short, stocky young man invested in a black wool suit, a string tie, and a black broad-brimmed hat. Hardeman writes that his outfit, combined with his naturally grave-looking countenance, made him look like an apprentice undertaker (and a sweaty one at that).[6]

At some point the two Sams decided they could cover more territory and give voters a clearer sense of their respective views if they campaigned together. Sharing a one-horse buggy day after day, they'd pull into a one-horse town, gather a crowd, and speechify. Standing above the curious on the back of their buggy, one or the other candidate would start in to speaking and then step down for his opponent's remarks.

As the campaign wore on and the candidates got better acquainted on their long rides through the countryside, they became close friends. "They began praising each other in such lavish terms," Hardeman and Bacon write, "that voters often were puzzled as to why such good friends were competing for the same job."[7] In one town, Gardner got sick and had to spend three days in bed. Although the man from Honey Grove seemed to be in the lead at the time, Rayburn refused to take advantage of his opponent's misfortune. He broke off campaigning and helped nurse him back to health. Once Gardner got to feeling better, the two resumed campaigning, still together.

Election day was Saturday, but it took a while for the ballot boxes from out in the county to be brought into the county clerk's office in Bonham. It took a while longer for the ballots to be counted by hand. The results would not be announced until the following Tuesday.

Anxious members of the Rayburn clan gathered at the family farm. Sam had arranged for a friend with a fast horse to be at the courthouse when results were announced and then ride the eleven miles out to the farm with the news. During the day, the Rayburns heard rumors that their man was winning big in the Bonham-Windom area, where the family was well known, but that Gardner was running up wide margins in his part of the county. They waited.

Along about sundown, they heard the hoofbeats. Barging through the front door seconds later, the friend was shouting, "You won, Sam! You won!"

By 163 votes.[8]

On January 8, 1907, two days after his twenty-fifth birthday, the youngest man ever elected to the legislature from Fannin County raised his farm-calloused hand on the floor of the Texas House to take the oath of office as a representative of the people. That winter day in Austin marked the beginning of the longest lawmaking career in American history, a career that had more impact on the electric cooperative movement than that of almost anyone short of Franklin Delano Roosevelt.

Rayburn the politician wasn't perfect. He could be short tempered, tough on his staff, and blind to emerging issues like racial justice, women's rights, and the environmental movement, and yet from the beginning until the end of his distinguished career, in Austin and in Washington, he never forgot where he came from or why he was there. Instead of living in the fancy Driskill Hotel during the legislative session, like other lawmakers, or even having a drink in the lobby bar, he shared a small room at a cheap boardinghouse. He was determined to save his five-dollar-per-day salary for law school tuition at the University of Texas.

Rayburn took classes for seven months before running out of money. He finished his legal training in the office of two experienced lawyers in Bonham and then joined their firm as a partner.

Shortly after Rayburn began his practice, one of his partners handed him the largest check he had ever seen, his share of the firm's monthly retainer from the Santa Fe Railroad. Rayburn handed it back. As a member of the legislature, he represented the people, not the railroad, he told his law partners. Lawmakers also received free railroad passes, but Rayburn returned his, even though he was homesick and lonely in Austin but too poor to go home for a visit.

During Rayburn's three terms in the Texas legislature, lawmakers reflected the progressive populist inclinations sweeping the nation at the time. They established groundbreaking regulations to rein in powerful banking and insurance interests, enacted a pure food law, and created an office to aid farmers. Rayburn voted for all these bills, as well as legislation establishing a maximum fourteen-hour workday for railroad employees and an eight-hour workday for telephone operators and railroad telegraphers. Lawmakers strengthened the state's antitrust laws, making them the strictest in the nation. In 1909, they established a fund to insure bank deposits, twenty-four years before the Federal Deposit Insurance program was enacted as part of Roosevelt's (and Rayburn's) New Deal.[9]

During his third term in the Texas House, the twenty-nine-year-old lawmaker was elected speaker, the youngest in the history of the Texas House. Despite the power and influence he had amassed in such a short time, he didn't tarry long in Austin. He was elected to Congress in 1912. In Washington, the man who had called a halt to campaigning to care for a sick opponent continued to focus on the needs of constituents, not himself. "Men who represent the people should be as far removed as possible from concerns whose interests he is liable to be called to legislate on," he said.[10]

Deep into his political career, when the bald head and permanent scowl were recognizable nationwide, Rayburn not only stayed connected to the people he represented; he also never lost sight of the fact that he served them, not the other way around. In a Kaufman County hamlet called Gastonia, Ray Patton's general store was for years the local gathering place, the unofficial community center. Rayburn dropped by whenever he was in the area and would often sit around the store for a couple of hours.

"He'd come out to my old store, and he always had some congressman or big-shot government official with him," Patton told Neal Johnson of Kaufman County Electric Cooperative years later. "He'd sit down, eat cheese and crackers and visit with all the folks as they'd come and go." It was Mr. Sam's way of staying in touch, Patton explained. "By the way," he added, "he always paid for his cheese and crackers."[11]

For the first twenty years of his congressional career, the stocky, hard-working son of Fannin County farmers, his hairless head as rounded as the capitol dome, was mastering the arcane ways of the House of Representatives. He chaired powerful committees, sponsored significant legislation, inspired confidence among his colleagues, and, ever ambitious, continued hoping he might yet reach his goal of becoming speaker.

When Franklin D. Roosevelt swept into office intent on rescuing the nation from the debilitating Great Depression, Rayburn was approaching the apex of his power and influence. His fellow Texan John Nance "Cactus Jack" Garner had become vice president, leaving the man from Fannin County the senior member of the powerful Texas delegation. He chaired the Interstate and Foreign Commerce Committee, one of five Texans heading important House committees. He and the new president had both turned fifty, and, as Rayburn himself noted, "Some men apparently ripen earlier than others and burn out early. I know that the period after age 50 was the best for me." His constituents seemed to agree.[12]

On the cold March afternoon that Roosevelt was sworn into office, fifteen million Americans, more than 25 percent of the workforce, were unemployed. Five thousand banks with total deposits of $3.6 billion had shuttered their doors, and one hundred thousand businesses were bankrupt. Millions saw their lifelong savings wiped out. One out of every four farms had been sold for debt and taxes. The nation was fearful, without hope.

Roosevelt vowed to do all in his power to turn things around, and Rayburn was eager to help him. "There must be an end to a conduct in banking and in business which too often has given to a sacred trust the likeness of callous and selfish wrongdoing," the president said in his inaugural address. "There must be strict supervision of all banking and credits and investments; there must be an end to speculation with other people's money."[13]

Less than a month after taking office, the new president sent Congress a comprehensive plan for strict federal supervision of investment securities sold in interstate commerce. Rayburn introduced draft legislation, labeled the Securities Act of 1933; it was referred to the committee he chaired.

That bill was the first of several pieces of legislation vital to the New Deal that Rayburn would shepherd through the House, usually having to wage bitter, grueling battles that took months, occasionally years, before he and the president prevailed. During the early years of the New Deal, he was deeply involved in legislation calling for the supervision of stocks and bonds, the breakup of railroad holding companies, regulation of the stock market itself, creation of the Federal Communications Commission, federal controls over oil production, and, in what would come to be called "the greatest congressional battle in history," the breakup of the investor-owned public utility holding companies.

Back home in North Texas, Rayburn's rural constituents were still without power and light. In fact, most of rural America was still in the dark, fifty years after electricity first illuminated cities and towns. Rayburn was sure he knew why. Powerful holding company interests, implacable opponents of the New Deal, were more interested in gouging consumers with exorbitant rates and bilking investors with worthless stock than they were in serving the public. Rayburn called them "bloodsuckers."

Holding companies first came into existence in the late 1880s, becoming favorite vehicles for the growing gas and electric power industry. Labeled the glamor industry of the early twentieth century, the utilities began attracting speculators and shifty financial manipulators. The operating utilities and unwary customers were their prey.

The holding companies were essentially financial devices that allowed bankers, investors, and speculators to control the stock of a utility without worrying all that much about the end result. Holding companies often controlled other holding companies in what were essentially pyramid schemes, with each layer extracting profit from the company or companies below. One holding company in the Midwest controlled 152 subsidiaries, which in turn controlled an additional 74 companies. Operating profits were of secondary concern; they were interested in quick returns from stock manipulation.

At the base of the pyramid were the utility operators generating and distributing gas and electric power. Under the thumb of the holding companies, their job was not to serve the public but to generate handsome profits from service revenues, users of electricity be damned. The arrangement translated to higher electric rates for customers, if they could get electricity at all. No federal regulation monitored either the industry or the generation of electricity by the private utilities.

The New Dealers agreed that the holding companies were parasites; they disagreed on the solution: regulate or abolish. Roosevelt seemed undecided.

Although Rayburn had tried to avoid "punitive legislation" during his career, he eventually concluded there was no satisfactory alternative to breaking the "bloodsuckers" short of abolition. "We must discourage this cancerous growth on our body politic and remove it," he declared. "If left alone it will jeopardize our financial institutions and perhaps destroy the Republic. The abuses of holding companies are indeed a major influence that brought on the Great Depression."[14]

In 1935, the president asked Rayburn to sponsor legislation toward that end. Senator Burton K. Wheeler of Montana, a venerable populist and chair of the Senate Commerce Committee, carried the companion bill. The battle was engaged. Wheeler, who was US senator Robert La Follette's 1924 running mate on the Progressive Party ticket—they polled nearly five million votes—told oral historian Studs Terkel years later that the battle was hard fought.

"These utility people are gonna destroy anyone who gets in their way," he recalled being warned by a friend. A few days later, two of their chief lobbyists dropped by his office to see him.

> I said, "Have you got any guns?"
> They said, "No, your boy frisked us."
> I said, "I told him to."
> They said, "How much time you gonna give us?"
> "One week."
> "We've got to have a month."
> "I'll give you a week, and the Government a week. If you can't tell what's wrong with the bill in a week's time, that's too bad."
> They said, "You're pretty cocky."
> "No," I said, "I'm just tired of crooked lobbyists and crooked lawyers."[15]

Relying on its enormous political and financial resources, the utility industry wasn't totally reliant on "crooked lobbyists and crooked lawyers." The industry mobilized banks, insurance companies, wealthy investors, and industrialists in a fight "to save free enterprise." The Rayburn-Wheeler legislation targeted the traditional American corporate structure, the industry warned, but its real victims would be the millions of ordinary stock and bond holders who relied on their dividends. They were invariably described as "widows and orphans."

A. J. Duncan, the president of the Texas Electric Service Company, declared that the Rayburn-Wheeler bill would "take from Texas rights for which Texas has shed blood; it would destroy the investments of literally thousands of people who have attempted to provide for old age by investing in the securities of this and other companies; it would largely destroy the fundamental principles of states' rights."[16]

Rayburn fought back. In a radio interview, he noted that the nation's two thousand utility operating companies, worth $20 billion, were controlled by about fifty holding companies. One banking house controlled more than a fourth of the nation's electric companies. "Instead of taking power, authority and management away from the local communities," the Texas congressman shouted on the House floor, "I want to take it away from New York, the Insulls in Chicago, and give it back to the communities of this country where it belongs."[17]

Biographers Hardeman and Bacon noted that Rayburn had a gift for translating esoteric policy issues into everyday terms. "The holding device is so clever," the congressman told his radio audience, "that a schoolgirl cannot use her curling iron, a housewife cannot clean her rug with a vacuum cleaner or preserve her food in an electric refrigerator, a schoolboy cannot turn on a light to read his lesson, a cook cannot light the gas in her stove without paying tribute to a holding company. It is included in the rate paid for the lights and gas."[18]

Lead-off witness for the utilities before the Rayburn committee was Wendell L. Wilkie, president of a holding company called Commonwealth & Southern Corporation and a future Republican presidential nominee. Handsome, self-assured, and eloquent, Wilkie didn't sit at a witness table humbly waiting to answer questions from committee members; he strode back and forth before the committee dais, making the case for the utilities and setting the tone of the discussion before lawmakers had a chance to. He conceded that the utility industry had been guilty of grave abuses, but those abuses were no reason to destroy the holding companies. They were necessary to provide financing for the operating companies, he argued. They were necessary to make the power industry more efficient.

Wilkie was a spellbinding witness for the holding companies, but, as newly elected Texas congressman Bob Poage recalled years later, "He would have greatly hampered the development of rural electrification had he been elected president. It would have certainly slowed down to a snail's pace."[19] Wilkie did not impress the bald, glowering Texan who chaired the committee. "I always thought he was kind of a smart-ass," Rayburn said years later.[20]

The holding-company battle lasted two hundred days, from the fall of 1934 into a sweltering DC summer. Claims and counterclaims, attacks and counterattacks sapped energy and left weary, short-tempered lawmakers at an impasse. On August 24, the man who years earlier had dragged a heavy cotton sack up and down endless rows until the job was done finally found a way to end the stalemate. He was able to forge a compromise that essentially eliminated the holding companies, with exceptions that the Securities and Exchange Commission could in theory grant. The battle was over, victory for the New Deal ensured.

As the first session of the Seventy-Fourth Congress drew to a close, columnists Drew Pearson and Robert S. Allen described one of the year's memorable scenes: "Representative Sam Rayburn, floor leader for the holding-company bill, being thunderously applauded by Republicans as well as Democrats for his fairness and patience in the utility fight."[21]

In a speech delivered on NBC in August 1935, the same week Roosevelt signed the bill, Rayburn observed that the president's signature "ended the hardest battle I have seen in more than 20 years in Congress. It was a battle against the biggest and boldest, the richest, and most ruthless lobby Congress has ever known."

Biographers Hardeman and Bacon observed: "Strange are the ways of politics—a man who owned neither stocks nor bonds, who for the first fifty years of his life neither understood nor cared about the intricacies of Wall Street, was called upon to lead the fight for the laws that sought to regulate the activities of high finance as never before in American history. It had been a battle of unprecedented intensity between big money and big government. Years later, Rayburn reduced the struggle to stark simplicity: 'Just call it the fight for economic justice.'"[22]

With the parasitic holding companies tamed, Rayburn could turn his attention to helping the people he knew best, his state's "unwilling servants of the washtub and water pump." He pushed rural electrification for them and for the common people of America.

3

THEY HAD "NEVER SEEN A MOTHER AND SISTER OVER A WASHTUB AS I HAD"

The plight of the power-deprived farmer was well known in Washington and in state capitals around the country long before Roosevelt was elected president. In 1909, FDR's distant cousin President Theodore Roosevelt appointed the Commission on Country Life to assess the growing disparities between rural and urban America. The commissioners included the man who proposed the study, Gifford Pinchot, the nation's chief forester, later recognized as the father of the conservation movement. Highlighting the lack of electric power in rural areas, the commission proposed a variety of methods to get power to rural Americans, including through federal hydroelectric power and the organization of electric cooperatives. Nothing came of the report.[1]

In 1923, the National Electric Light Association, forerunner to the Edison Electric Institute, created the Committee on the Relation of Electricity to Agriculture (CREA) to study whether rural Americans would even use electricity if they had it. The CREA, financed by the private utility industry, government agencies, and electric equipment manufacturers, set up demonstration projects around the country.[2]

The most notable of these projects was near Red Wing, Minnesota, a dairy-farming community along the Mississippi River southeast of St. Paul. The project, under the auspices of the University of Minnesota School of Agriculture, strung about six miles of line to serve between eighteen and twenty farmhouses. Ten of the farms were equipped with every type of electric appliance imaginable, including lights in barns, milk sheds, and chicken coops; electric motors for cutting wood; electric pumps for irrigation and water for home use; and electric dryers for dehydrating hay.[3]

"Exhaustive experiments and research will be conducted," the *Minneapolis Daily Star* reported. "Investigators, for example, will compare the growth of beef cattle, watered by electricity, with herds watered from barnyard troughs. The increased income from electrically lighted chicken houses will be noted. Experiments may be made by plowing by electricity, supplied to motor-driven plows either by storage battery or cables."[4]

Agricultural economists and professors from around the country dropped by to observe. Newspapers paid attention. An editorial writer for the *Daily Star* observed that the participants in the experiment had "a fair chance of rendering one of the greatest services to agriculture—comparable to the invention of the reaper or the cotton gin." The editorial writer was skeptical it would happen. Noting that nearby Ontario, Canada, was relying on government subsidies to wire its farms, he said he wasn't sure such an approach would work in the United States. "The heart of the problem," the writer concluded, "is to develop such an even maximum load per farm as will pay the farmer and the agency furnishing the power whether it be a private company or a public project."[5]

"The farmers in the Red Wing experiment discovered that they and the power companies got a greater return the more electricity each farm employed," Marquis Childs reported decades later. "The rate of return went up steadily as more and more uses were made of the current that had been brought from the city."[6]

Despite those findings, the leaders of the private power industry continued to ignore rural America. Even after reading the reports of the commission they had created, they concluded that the farmer was not worth their time and effort. They continued to insist that rural areas were too limited to warrant the investment.

For the farmer, dealing with the power companies was infuriating, humiliating. LBJ biographer Caro writes:

> For two decades and more, in all states of the country, delegations of farmers, dressed in Sunday shirts washed by hand and ironed by sad iron, had come, hats literally in hand, to the paneled offices of utility-company executives to ask to be allowed to enter the age of

> electricity. . . . But in delegations or alone, the answer they received was almost invariably the same: that it was too expensive—as much as $5,000 per mile, the utilities said—to build lines to individual farms; that even if the lines were built, farmers would use little electricity because they couldn't afford to buy electrical appliances; that farmers wouldn't even be able to pay their monthly electricity bills, since, due to low usage, farm rates would have to be higher—more than two times higher, in fact—than rates in urban areas.[7]

In 1935, nearly a half century after urban America had started to rely on electricity, more than 6 million of the nation's 6.8 million farm families were living in darkness. Thanks in large part to the recalcitrance of the powerful private-power industry, America had become, in essence, two nations.[8]

"When they were told that perhaps they had a social responsibility to bring electricity to the farms," Childs writes, "power company spokesmen retorted tartly that the utility industry was not a charitable institution. If the farmer wanted electricity on his farm, said the power companies, or most of them, then he must make sacrifices in order to pay for it. As CREA had shown, only a few farmers could afford to pay the two or three or five or six thousand dollars that it cost to get 'hooked up.'"[9]

"But fairness—or social conscience—was not the operative criterion for the utilities; their criterion was rate of return on investment," Caro writes. "As long as the rate was higher in the cities, they felt, why bother with the farms? Their attitude was reinforced not only by their political power but by their contempt for country people."[10]

Their attitude was gratingly evident in Johnson's Hill Country district, where Texas Power & Light condescended to allow farms within fifty yards of the one TP&L line running into the countryside to hook up to it. A number of farmhouses a few yards beyond the limit were out of luck, even when the farmer offered to pay the additional cost. Can't do it, the company explained; to make an exception for one farmer might result in exceptions for many farmers. "Farmers whose homes were just beyond the fifty-yard limit, farmers who could see those lines every day of their lives, were unable to use them," Caro writes, "while they had to watch their wives year by year slaving at tasks that electricity would have made so much easier."[11] Some farmers told the power company they would be willing to move their houses so they could be within the fifty-yard requirement. Can't do it, a company spokesperson explained. It would set a precedent.

The farmer and his champions weren't giving up. In the years following the Great War, Pinchot resumed efforts to bring power to rural Americans. Elected governor of Pennsylvania, he appointed an engineer named Morris Llewellyn Cooke to direct his Giant Power Plan to electrify rural Pennsylvania. "From the power field perhaps more than any other quarter," Pinchot wrote, "we can expect in the near future the most substantial aid in raising the standards of living, in eliminating the physical drudgery of life, and in winning the age-old struggle against poverty. . . . Our first concern must be with the small user . . . particularly the farmer."[12]

Cooke proposed building generating plants near the Pennsylvania coalfields and using the abundant supplies of power these plants would produce to provide farmers with affordable electric energy. The Keystone State's conservative legislature rejected the plan, although it attracted the attention of the governor of Pennsylvania's neighboring state, a man named Franklin D. Roosevelt.[13]

In 1931, Roosevelt hired Cooke as a consulting member of the New York Power Authority and tapped him to investigate whether power developed on the St. Lawrence River could be distributed to farmers and small consumers. Cooke's study team, made up mainly of experts with whom he had worked in Pennsylvania, determined that the power companies were wrong when they insisted that it would cost about $2,000 to build a mile of line. Cooke's group determined that it could be done for between $300 and $1,500 per mile less than the power companies claimed. The study convinced Cooke that rural electrification was indeed feasible, regardless of the power companies' claims.[14]

In 1933, the newly elected President Roosevelt hired Cooke to head the Mississippi Valley Corporation, with specific responsibilities to prevent flooding along the mighty Mississippi and create economic development. Cooke used the assignment to make the case for a national rural electrification plan. He called his report to the president "This Report Can Be Read in 12 Minutes." In those dozen minutes, he noted that as the nation emerged from the Depression, the federal government should not only contribute to the nation's social life but also require united investment beyond the interest and capacity of the private power industry. Rural electrification, he contended, met both those requirements.[15]

Cooke noted that more than five million farms in the United States were "entirely without electric service," and private utilities wouldn't take on the problem. "Only under Government leadership and control is any considerable electrification of 'dirt farms' possible." In "How [to] Make the Start," one of the shortest of the memo's fourteen sections, he said the government could spend $25,000 or $50,000 on a survey, but a $100 million investment "actually to build . . . rural projects would exert a mighty influence." And this could be done without stepping on the toes of established utilities. "This proposal does not involve competition with private interests," he pointed out. This plan calls for entering territory not now occupied and not likely to be occupied to any considerable intent by the private utilities."[16]

The next six sections—"Source of Power for Rural Services," "Distribution Lines," "Large Use the Key to Lower Rates," "Rates," "Distance between Farms," and "Financing of Lines"—are the heart of Cooke's proposal. They make the economic case for federal involvement. Cooke argued that sufficient power supply could come from contracts with private utilities, local diesel plants, or hydropower. "The electric current itself [retail], in any case, can be made available to the rural population at a figure considerably below what is charged for it on existing rural lines." Distribution lines could be built for $500 to $800 a mile, and this cost could be amortized in twenty years at an interest rate of 4 percent. He did not say much about transmission. In "Large Use the Key to Lower Rates," he asserted that farmers "must learn to substitute [electricity] for human labor" to make rural electrification economically viable. Otherwise, high rates would be a disincentive to electrify.[17]

Cooke's proposal also established the consumer-density standard of three consumers per mile of line—REA loan applications that did not meet this minimum were not approved—and predicted the emergence of consumer-owned local providers of electricity: "farmers' mutuals operating without profit." These homegrown utilities "should receive Federal and/or State aid in the form of expert engineering, accounting and management advice, as farmers are now advised by experts in farm management." He also envisioned the REA itself and its loan program. Such a federal agency would employ "socially minded electrical engineers, who, having standardized rural electrification equipment, will cooperate with groups within the several states in planning appropriate developments." Where federal financing was needed it would be self-liquidating.[18]

"The '12-Minute Memo' was the document that convinced Ickes and FDR of the desirability and 'do-ability' of rural electrification," the National Rural Electric Cooperative Association (NRECA) said in its 1984 photo book about rural electrification's first fifty years, *The Next Greatest Thing*. It was the basis for rural electrification.[19]

"The REA was fundamentally different from much of the work done by the government during the New Deal. Instead of direct government intervention, the REA's founders chose to invest almost exclusively in capital, instead of labor," Noah Karr Kaitin, a business student in Cornell's School of Industrial and Labor Relations, wrote in 2013. As Kaitin noted, the REA's founders did not intend for government to control the rural utility market. Their aim was to help set up cooperative enterprises—electric co-ops—and then provide them the capital, the know-how, and the tools they would need to get power to rural America. "By choosing this path," Kaitin wrote, "the REA was able to closely oversee the development of the cooperatives, without needing to manage every bit of the work on the ground. Had the REA chosen to simply fund the existing private utilities, rural electrification would have likely proven a far costlier enterprise."[20]

Roosevelt hardly needed convincing, as he reminded an audience of some forty thousand people in Barnesville, Georgia, in August 1938. Not unlike his friend Rayburn, FDR had the ability to personalize dry public policy. He said:

> Fourteen years ago, a Democratic Yankee came to a neighboring county in your state in search of a pool of warm water wherein he might swim his way back to health. . . . There was only one discordant note in that first stay of mine at Warm Springs. When the first-of-the-month bill came in for electric light for my little cottage, I found that the charge was 18 cents a kilowatt hour—about four times what I pay at Hyde Park, New York. That started my long study of public utility charges for electric current and the whole subject of getting electricity into farm homes. . . . So it can be said that a little cottage at Warm Springs, Georgia, was the birthplace of the Rural Electrification Administration."[21]
>
> In Roosevelt's view, the REA wasn't just an effort to improve farmers' lives; it was an effort to kick-start economic recovery. With high unemployment and power companies unwilling to create a power infrastructure in rural areas, rural Americans were, in effect, shut out of the American economy. They could not participate in the nation's recovery.[22]

The president had signed an executive order creating the REA three years earlier, on May 11, 1935. With Cooke in charge, the new agency set up shop in the stately old Westinghouse Mansion near Dupont Circle.

As part of the Emergency Relief Appropriation Act of 1935, the REA had the primary purpose, initially, of providing work for the jobless, but it quickly became obvious to Cooke and his team that it wasn't feasible to hire unskilled workers to plan and build rural electric lines. Even if they could be quickly trained, the newly hired workers would be working for private utility companies. From the beginning, the companies showed almost no interest in rural service, even though the federal government offered grants and subsidies. The private companies continued to insist that the whole idea of rural electrification made no sense.[23]

Nine days after Roosevelt's executive order, executives from fifteen of the nation's largest power companies convened a meeting at the Lafayette Hotel in Washington to craft a response to the order. The executives formed a committee, and the committee issued a report in a fourteen-page letter. The conclusion was that "there are very few farms requiring electricity for major farm operations that are not now served" and that "additional rural customers must largely be those who use electricity for household purposes."[24]

Cooke now knew exactly where he stood. He began exploring alternatives. He first considered municipalities, aware from his own experience that cities large and small owned their electrical systems. Los Angeles, for example, had been furnishing power for years to farmers and fruit growers at city rates. A ruling by the Missouri Supreme Court preventing municipalities from extending their lines beyond their corporate limits complicated that plan. The court held that the farmer could come to the city limits and tap into power from a municipal plant, but the city could not go beyond its limits to the farmer.[25]

Farmers were getting restless. Marquis Childs reported that letters were pouring into Washington from farmers who had heard something about a new program and wanted to know more. A rancher in Colorado wrote that he had read "where a man had been appointed to have charge of a committee or bureau where the President wished to have farms electrified." Where could he find out about it? An Indiana farmer wrote to the president that he was so interested in getting power that he climbed aboard his "good black saddle-horse" and rode around to neighboring farms. Along with his letter, he sent a petition with forty-six names, asking the president for help.[26]

Cooke responded: "Find out how many farmers living within, say, five or ten miles of your home in any direction would pay for electricity if they could get it at a moderate price. Ask them about how much they would use—and for what purpose—grinding feed, heating water, preserving fruit in an electric refrigerator and, of course, lighting their houses and pumping water. When you have the facts, send them to us."[27]

Whether the Indiana farmer complied is not known, but with private utilities and municipalities pretty much out of the picture, it was becoming obvious how the REA might work. Cooke reverted to an idea with venerable roots in Merry Olde England.

In the early decades of the nineteenth century, the Industrial Revolution transformed traditional ways of life in the north of England. As David J. Thompson points out in his history of the modern cooperative movement, a rural nation was transformed almost overnight into an urban society. Prosperity came to Britain, but it wasn't shared equally. Social dislocation was widespread—and devastating. The city of Manchester, Britain's emerging industrial behemoth, grew from eighty-five thousand people in 1801 to four hundred thousand in 1851. As rushing streams of the Pennine Hills brought cheap power to the small mill towns in the vicinity, Manchester became "cottontopolis." The city and environs shipped processed cotton and wool cloth to the rest of the world, even as hundreds of thousands of rural and small-town people lost their work and their way of life when power looms replaced hand looms and factory owners consolidated production. "Conditions were ripe for revolt," Thompson writes. "Unemployment, near starvation, the poorhouse, disease and epidemics, child labor—all these were the lot of working people. The machinery went ever faster, the pay was never enough, the price of everything went up, and food was always too expensive."[28]

In 1811–12, an organizer named Ned Ludd persuaded weavers in Nottinghamshire, South Yorkshire, and Lancashire to band together and stand up for themselves. The Luddites, as they were called, intended to hack to pieces the gig mills and wide knitting frames that were destroying their livelihoods. Twelve thousand soldiers put down the riots (although the label lived on).[29]

In April 1825, handloom weavers rioted across central and eastern Lancashire. Roaming from village to village, bands of weavers broke into factories and tore apart more than a thousand power looms in four days. Again, soldiers rushed in to put down the rebellion. Thirty rioters went to prison; another ten were shipped off to faraway Australia.[30]

In the summer of 1843, a young writer living in London journeyed to the north of England to observe conditions among working men and women (and children). In a speech to the Manchester Athenaeum, a local institution dedicated to providing workers an education, the author urged employers and employees to come together and share "a mutual duty and responsibility." A week later, thirty-one-year-old Charles Dickens returned to London to begin work on *A Christmas Carol*.[31]

In 1844, a German-born businessman living in Manchester proposed a more radical solution to the plight of the displaced poor in the north of England. Friedrich Engels wrote *The Conditions of the Working Class in England*, a book that came to the attention of Karl Marx and formed the basis of *The Communist Manifesto*, coauthored by both men and published in 1848.[32]

Conditions in the small Lancashire town of Rochdale, ten miles northeast of Manchester, were typical. In the early years of the Industrial Revolution, the Rochdale Canal was one of the major navigable canals of the United Kingdom, used for hauling coal, cotton, and wool. Rochdale itself was famous for its flannels. (The name is pronounced RAHCH-dale.)

As textile production became mechanized, local weavers spiraled into poverty. Thompson quotes a contemporary reformer who reported on the appalling conditions: "The employment of women at once breaks up the family; for when the wife spends twelve or thirteen hours every day in the mill, and the husband works the same length of time there or elsewhere, what becomes of the children? They grow up like wild weeds; they are put out to nurse for a shilling or eighteen pence a week, and how they are treated may be imagined. Hence the accidents to which little children fall victims to a terrible extent." Thompson also quotes the famed children's writer Beatrix Potter, who in 1892 reported to the Cooperative Congress on the dire condition of Rochdale children in 1870. Their life was difficult from the beginning, Potter reported, because their mothers went to work in the mills while their children were still in the cradle. She reported that at ten years of age, a Rochdale child was one and a half inches shorter and five pounds lighter than the average British schoolchild. Rochdale children went to work in the mills at age ten instead of the standard age of twelve.[33]

Workers pleaded with Parliament and the Crown, to no avail. The industrialists, despite the admonition from Dickens and others, ignored the plight of those their factories had displaced. Desperate, the Rochdale textile workers resolved to take matters into their own hands, not by force of arms or strikes but through mutual aid. On October 24, 1844, twenty-eight weavers and skilled artisans formed the Rochdale Society of Equitable Pioneers, a cooperative effort to pool their resources and purchase needed supplies in larger volumes and at lower prices. Theirs was primarily an effort to escape the thrall of factory- and mill-owned shops that regularly overcharged for low-quality merchandise and adulterated food.

The Rochdale Society of Equitable Pioneers came to realize that cooperation took many forms. Instead of each family making the long trek into the city for supplies, for example, they took turns, with one of their number filling orders for everyone. They pooled resources to take advantage of volume buying. The small first-floor storefront the group rented on warehouse-lined Toad Lane as its base of operations is considered the birthplace of modern cooperative businesses.[34]

The Toad Lane store was nothing to brag about when it opened in 1844. Tradespeople and storekeepers gathered outside on a raw December night, curious to see what a group of poor weavers were offering. When the doors opened and they sauntered in, they began laughing and jeering. They saw two rude plank tables holding a meager stock of goods: twenty-five pounds of butter, fifty-six pounds of sugar, a sack of oatmeal, and a few barrels of flour. Two tallow candles lighting the shop were meant to be for sale, but they had to be put to use on the spot when the gas company, skeptical of the weavers' venture, refused to supply the shop.[35]

"Although the weavers had little to sell that night, their cooperative effort eventually became so successful that it gave birth to modern cooperatives that include housing co-ops, credit unions and farm marketing and rural electric co-ops," *Texas Co-op Power* noted in 1944.[36] A 1985 *Texas Co-op Power* article stated, "The idea is alive today in 67 countries with more than 360 million co-op members."[37]

The Rochdale Pioneers did not originate the idea of individuals working cooperatively to achieve what they could not accomplish alone. The concept had been around for many years, in Europe and America. The Mayflower Compact, which the Pilgrims had signed before sailing to America, was a cooperative, as were the mutual fire insurance companies Benjamin Franklin had founded in colonial Philadelphia. The significance of the Rochdale Pioneers was not in what they did but in how they did it.

They built their organization on a set of practices and descriptions now known as the Rochdale Principles. None of these ideas was totally new, but together they produced a business system that was unique. The pioneers hoped to make a profit, of course, but only within certain parameters. They preached principles including democratic control, open membership, fixed or limited return on subscribed capital, dividends on purchases, trading strictly on a cash basis, and selling only pure and unadulterated goods. They believed in educational opportunities for members and political and religious neutrality.

"The primary goal of the Rochdale Pioneers was modest—to simply lower their production costs to increase their incomes," historian Thompson notes. "It is unlikely that any of these desperate weavers with little formal education cared much about the concept of cooperation or realized that they were establishing a long-term worldwide movement. While modern-day business practices have brought change to this original set of principles, the essence of the cooperative business model embodied in them continues to this day and are clearly the basis of the 8 Cooperative Principles that form the basis for all cooperatives formed since."[38]

The Rochdale Pioneers had established a set of principles, and yet European farmers had been "cooperating" for centuries. In America as well, farmers had established milk cooperatives, grain cooperatives, and other joint ventures in an effort to gain a bit of leverage over the banks, railroads, and private utility companies that kept them in thrall. Immigrant Scandinavian farmers in the Midwest and Czech, Bohemian, and Moravian farmers in Central Texas had organized co-ops in the old country.

Midwestern farmers even organized a small electric co-op in 1914, and yet most farmers found using cooperatives to draw electric power to their farms to be a bit mystifying. "Electricity wasn't like wheat or fertilizer that could be held in the hand and looked at," noted NRECA editors of *The Next Greatest Thing: 50 Years of Rural Electrification in America*. "It came from far away over humming lines. And you needed engineers and lawyers to tame it."[39]

By the mid-1920s, the rest of the Western world was far ahead of the United States in efforts to provide electricity to rural areas. Rural Sweden was 50 percent electrified; France, 71 percent; Finland, 40 percent; Denmark, 50 percent; and Czechoslovakia, 70 percent. The European countries obviously felt the weight of social responsibility to extend electricity to their rural areas, but America did not. As late as 1935, only one in ten rural Americans had electricity, and most of those lived near cities and towns. The 1920 census listed six and a half million farms in America.

On August 7, 1935, President Roosevelt issued Regulation Number 4, transforming the REA into a lending agency. "This was the first and probably the most far-reaching fundamental policy decision in the history of

REA," wrote H. S. Person in 1950. "It established promotion of rural electrification as an orderly lending program on an interest-bearing, self-liquidating basis. It made rural electrification a national business investment. It made possible the subsequent great achievement of REA."[40]

On January 6, 1936, Sam Rayburn in the House and George W. Norris of Nebraska in the Senate introduced the Rural Electrification Act. In years to come, the Texan would tell people that the law creating the REA, and legislation creating farm-to-market roads, were his proudest achievements. Both pieces of legislation, he would explain, helped "the real people."[41]

Congressman Wright Patman, Rayburn's friend and fellow populist Democrat, represented an adjacent North Texas district. Patman, who was about as folksy as Rayburn, offered a unique way of explaining what his friend had accomplished with the bills creating farm-to-market roads and the REA. He observed in a 1996 oral history interview at the LBJ Presidential Library, that those "two great programs" meant that

> a person could build a home out on a little farm-to-market road, and he would have every accommodation that you could get at the Waldorf-Astoria Hotel in New York. You could have a radio, you could have lights, you could have refrigeration, and you could have everything that they had in the Waldorf-Astoria with a good highway, a good road right in front of your house, with delivery service for newspapers and everything else. In addition to that, of course, one living there would have the benefit of the Waldorf-Astoria tenants, because they would have a place to park and free air to breathe.

"And at a little less price per night," Patman's interviewer, historian Joe Frantz, added.[42]

The House version of the REA bill went to the powerful Interstate and Foreign Commerce Committee, chaired by Rayburn. "And he simply had enough influence that they referred the legislation to his committee, although it probably never belonged there," longtime Texas congressman Bob Poage recalled years later. "And it has consistently been before Agriculture since I've been in Congress."[43]

The Senate bill went to the Agriculture Committee, where Norris was the ranking Republican. The Nebraska senator, proud to be known as the father of the Tennessee Valley Authority (TVA), was as deeply committed to rural electrification as his House colleague from Texas. He served under seven presidents, from Theodore Roosevelt to Franklin Roosevelt. He and Rayburn had served in Congress together for twenty-four years. According to biographers Hardeman and Bacon, Rayburn considered Norris to be "the most practical progressive with whom he ever served—a rare individual who not only aroused the citizenry with a new idea but one who developed the skill to write a bill and pilot it through hazardous legislative shoals to become the law of the land."[44]

Like Rayburn, the Nebraskan also was familiar with the cavernous gap between urban America and rural. First elected to the House of Representatives in 1908, he spent much of the 1920s building support for the TVA, a massive government project that would build a series of twenty-one dams on the Tennessee River to provide, among other things, cheap electricity to tens of thousands of farm families. He finally succeeded in 1933. The TVA was a forerunner to the REA. As a reward for his efforts, Norris was forced out of the GOP by the pro–big business faction of the party and had to run for reelection as an independent in 1936.

The Norris-Rayburn legislation completely revised the basic thrust of Roosevelt's executive order creating the REA. Instead of relying on private companies or cities to build lines into the countryside, farmers would band together cooperatively. The REA would become a government lending institution, providing several types of loans. The first would be used for building power lines and for generating and transmitting electricity. Another type of loan would be available for farmers to purchase electrical equipment, including wiring for the farmhouse, water system installation, and other labor-saving devices around the farm and in the kitchen. The co-ops would borrow the money, build and own the lines, and repay their federal loans through profits from the sale of electricity to farmers. Those "sad irons" would become paperweights or doorstops, never-to-be-missed mementos of an increasingly distant way of life.

Rayburn reminded his House colleagues about the need. "In Collin County there are 6,069 farms," he said in opening remarks on the House floor, "and only 83 have electricity. In Fannin County, 5,894 farms and 94 electrified; in Grayson County 5,169 farms, 101 electrified. In Hunt [County] 5,905 farms, 219 electrified; in Kaufman [County] 5,131 farms, 88 electrified; in Rains [County] 1,691 farms, 4 electrified; in Rockwall [County] 1,031 farms, 32 electrified."

"Primarily, the aim is to put an end to needless drudgery on the farm and to make possible a new and far higher standard of rural living," Rayburn said. "There is not an undertaking more important to the rehabilitation of our agriculture than rural electrification."

Another of Rayburn's fellow Texans, a fiery congressman from San Antonio (and future mayor of the city), tore into the private utilities. "So long as we look at the matter of rural electrification in the light of economic greed, private profit and speculation, nothing will ever be done for the farmer," Maury Maverick said.

The power companies didn't fight the REA's new approach, in part because they believed the program would fail. They were gravely mistaken, Rayburn said, having "never seen a mother and sister over a washtub as I had. They had never seen them heating irons in a fireplace; they had never stuck their hands in dirty coal-oil chimney lamps."[45]

As it turned out, Rayburn's leadership and Norris's political and legislative acumen were more than sufficient to the task. The House approved the measure without a record vote. The Senate passed its own version a few days later. After conference committee members resolved minor differences, a final bill went to the president for his signature on May 20, 1936. He signed it the next day.

Legislation opened the door, but a great many obstacles remained before rural America was empowered. As Marquis Childs pointed out, the REA had to devise legal, economic, and engineering criteria to determine whether a proposed cooperative could become a financially sound business. Different states had different requirements for setting up a co-op. In addition, co-ops would be operating in a cutthroat industry, so they had to be structured in such a way that the members wouldn't lose control of their enterprise to those who might destroy it. Engineers had to develop cheap and durable lines. Each co-op would have to find a capable manager who would function as both engineer and administrator. And, as Childs notes, managers would have to be smart enough and tough enough to negotiate with utilities providing wholesale power, "utilities that were often antagonistic."[46]

In 1937, the REA drafted the Electric Cooperative Corporation Act, a model state law for the formation and operation of rural electric cooperatives. By 1939, the REA had helped establish 417 electric co-ops, serving 288,000 households. All across the country, lights were coming on.

Like the Rochfield cooperative in England more than a century earlier, America's electric co-ops organized around a set of core beliefs, now known as the 7 Basic Principles:

1. Open and Voluntary Membership: Membership in a cooperative is open to all people who can reasonably use its services and stand willing to accept the responsibilities of membership, regardless of race, religion, gender, or economic circumstances.
2. Democratic Member Control: Cooperatives are democratic organizations controlled by their members, who actively participate in setting policies and making decisions. Representatives (directors/trustees) are elected among the membership and are accountable to them. In primary cooperatives, members have equal voting rights (one member, one vote); cooperatives at other levels are organized in a democratic manner.
3. Members' Economic Participation: Members contribute equitably to, and democratically control, the capital of their cooperative. At least part of that capital remains the common property of the cooperative. Members allocate surpluses for any or all of the following purposes: developing the cooperative; setting up reserves; benefiting members in proportion to their transactions with the cooperative; and supporting other activities approved by the membership.
4. Autonomy and Independence: Cooperatives are autonomous, self-help organizations controlled by their members. If they enter into agreements with other organizations, including governments, or raise capital from external sources, they do so on terms that ensure democratic control as well as their unique identity.
5. Education, Training, and Information: Education and training for members, elected representatives (directors/trustees), CEOs, and employees help them effectively contribute to the development of their cooperatives. Communications about the nature and benefits of cooperatives, particularly with the general public and opinion leaders, help boost cooperative understanding.
6. Cooperation among Cooperatives: By working together through local, national, regional, and international structures, cooperatives improve services, bolster local economies, and deal more effectively with social and community needs.
7. Concern for Community: Cooperatives work for the sustainable development of their communities through policies supported by the membership.

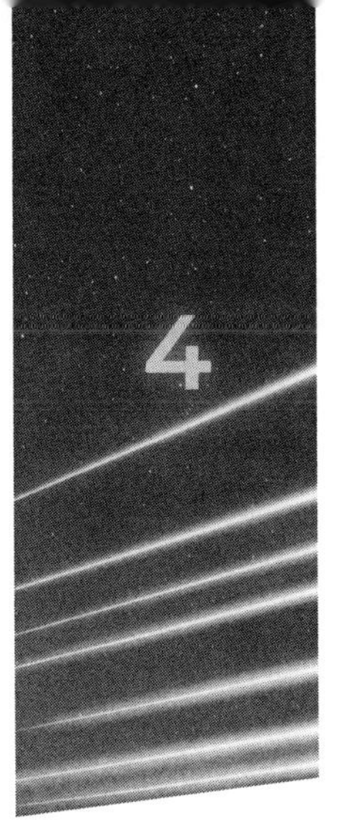

"THIS IS THE ONLY WAY TO GET THAT WOMAN OFF OUR NECKS"

On State Highway 95, between Taylor to the south and Temple to the north, are three small towns: Granger, Bartlett, and Holland. About six miles apart from each other, all three were settled by Czech and German immigrants in the 1870s and 1880s. Farmers in the old country, most of the newcomers landed at the Port of Galveston and made their way to Central Texas, drawn to the rich, rolling blackland soil that was perfect for growing cotton and corn.

In Granger, the settlers established a Czech-language newspaper, *Našinec*. For many years the only Czech paper in Texas, *Našinec* stayed in business for more than a century. Holland, named not for the country but for an early settler, was home to one of the first steam cotton gins in the area. Bartlett was the most prosperous of the three. When the Missouri, Kansas & Texas Railroad reached the town in 1882, Bartlett became a shipping point for cotton, grain, livestock, and produce. Within a few years, sturdy brick commercial buildings lined Clark Street downtown (and still do). They housed three banks, a hotel, a post office, a butcher shop, and a dry goods store. "The handsome brick buildings, occupants' names carved in stone above entryways, suggested permanence, respectability and confidence in the future, and so did Bartlett's newly constructed architecturally distinguished public school," Carol Moczygemba wrote in *Texas Co-op Power* in 1998.[1]

The city fathers of Bartlett, boasting a peak population of 2,200 in 1914, built a railroad of their own, running from Bartlett to the little town of Florence, twenty-five miles to the west. The Bartlett Western may have been an expression of local confidence in the town's continuing prosperity, but it was not particularly successful. Local wags called it the Bullfrog Line because it jumped the track so often. Others cracked that "BW" stood for "Best Walk."[2]

Bartlett residents, like farmers across the state, could not have predicted that the precipitous decline of cotton prices in the early 1920s—from $1.59 to $0.45 a bale—would not only cost the BW 25 percent of its business but also force the company into bankruptcy. As the Depression tightened its grip on rural America, Bartlett's population declined. Farmers gave up. Young people migrated to Austin or Dallas or Houston looking for work. Most never returned.

Perhaps Depression desperation was one reason those who stayed made history. In 1935, shortly after President Roosevelt signed the Rural Electrification Act, three area farmers put up fifty dollars each to create what they originally called the Bartlett Community Light & Power Company. "When we started, we had no idea what it would amount to," recalled W. R. Janke, the first president of the board of directors. The money, he said, "came out of our own pockets, and it was little enough to pay."[3]

The Bartlett farmers sent an engineer to Washington, DC, to investigate the possibilities of an REA loan. He came back with an optimistic report, so the farmers arranged a meeting with attorney J. B. Morris and decided to apply for a charter. "We had been trying to get a private power company to install electricity for us," Janke recalled nine years later. "They not only said they couldn't do it but that it couldn't be done at all."[4]

In November, the group received word from Washington that Bartlett had been approved for $33,000, one of the first ten loans the REA made. Their plan was to create a distribution system to serve nearby farms and contract with Bartlett's municipal utility to generate the electricity and provide other services, such as reading meters.[5]

The Bartlett farmers' plan to buy power from Bartlett's municipal generating plant, established in 1905, made sense. It allowed them to move ahead more quickly than other rural co-ops. On a cold November day, eight men—BCL&P volunteers and paid employees—gathered at a spot just outside town and began actual construction, and by nightfall they had performed a historic act. They had set the first section of thirty-six poles into the heavy black soil for a fifty-eight-mile line. The first REA-financed project in the nation, it's now the Bartlett Electric Cooperative.[6]

On March 7, 1936, a local farming couple, Charles and Lydia Saage (pronounced "soggy"), paid the five-dollar deposit for an electric meter and were given the honor of throwing the switch. The Saages became the first farm home in America to receive REA-financed power. The nearby Stokes family farm was the second. One hundred twenty Bartlett-area members soon joined them. "They were the first on the line going down that road out of Bartlett," said Mary Saage, a daughter-in-law, in a 2010 interview with *Texas Co-op Power*. "That was the reason they got electricity before anybody else." (Mary Saage, ninety at the time of the interview, recalled that her first summer job was working in the office for what would soon become Bartlett Electric Cooperative.)[7]

Bartlett was the first, and the rest of rural Texas would eventually follow, though not without misgivings. Texas stayed dark longer than most states, particularly in areas where low population density made even REA loans difficult to get. An innate stubborn skepticism on the part of Texas farmers was also a factor. They had been burned too often, they would tell you—by power companies and by the federal government.

For his history of the Kaufman County EC, Neil Johnson interviewed Jim Copeland, a teenager in the 1930s who worked in the Copeland family's general store in the Ola community of southeastern Kaufman County. On a summer morning in 1936, young Copeland witnessed the skepticism firsthand. He was busy cleaning black smudge out of the kerosene lantern globes they would use around the store after the sun set that evening, when his father stepped outside to listen to a man from Washington speaking to a crowd gathering in a nearby churchyard. "I'm going out to see what that fellow has to say," the elder Copeland told his son.[8] He came back to the store unimpressed. "That man is the biggest liar I've ever heard," he grumbled. He was talking about Sam Rayburn. Jim Copeland was surprised, he told Johnson more than sixty years later, since his dad rarely said anything bad about anybody. This time, though, the man simply couldn't believe what their congressman was saying—that soon electric lines would be lining the roads, bringing light and power to the Ola community.[9]

Rural Texans taking time off from their hard work and driving from one farmhouse to the next gradually bore fruit all over the state. Kaufman

County was one of the early co-ops organized. In the mid-1930s, 2,679 people lived in Kaufman itself, the county seat. Like the residents of Johnson City in the Hill Country, they got their electricity from Texas Power & Light, with power generated at the company's plant near the small town of Trinidad, on the Trinity River. Before TP&L, Trinidad got its electricity from Jay's Light Plant between 5:00 p.m. and 10:00 p.m. and also between 5:00 a.m. and 8:00 a.m. However modest, it was more than rural residents had. They were powerless until they got themselves organized in early 1938.[10]

Jim Copeland, whose storekeeper father had been an early skeptic, recalled years later that not long after the co-op was formed, residents of Ola and other small communities in the county began to realize how the arrival of electricity would change their lives. Community leaders knocked on doors, trying to generate as many memberships as possible. Some offered to help their neighbors pay the five-dollar membership fee. "If you didn't get enough people to sign up," Copeland told Neal Johnson in 1996, "you weren't gonna get electricity. It was just that simple, not hard to understand. So, we got out here and hit the road and knocked on doors."[11]

Their efforts came to fruition on November 15, 1938, when one hundred miles of line starting at a substation on State Highway 175 were energized. It served the Jiba, Becker, and Ola communities. "I was the proudest young man there ever was to see electricity come," Jim Copeland recalled, "'cause I didn't have to pump and clean those lanterns that we used in our store every night." Copeland also remembered that many of the store's regular customers had never seen electric gas pumps with numbers showing the gallons purchased and the total amount to be paid. "I will never forget one fellow who was watching the new electric pumps operate and turned to his father and said, 'Well, you finally got you some educated gas pumps.'"[12]

In West Texas, dive-bombing bugs were a problem before the advent of electricity. Mrs. C. W. Rose, a Swisher County public school teacher, recalled that it was customary to use kerosene lamps at home and Coleman gasoline lamps to light the school building. "Occasionally the mantles were knocked out by flying candle bugs," she recalled. "Before leaving the Elkins School, we had bought an Aladdin lamp. What an improvement that was!" she said, even though somebody always had to keep pumping them up.[13]

One of the myriad daily chores for the West Texas housewife was to keep the lamp globes clean and the wicks trimmed. "She had to watch constantly so the blaze didn't run down the wick and into the kerosene," Tulia resident

R. B. Dawson recalled. "When that happened, the entire lamp had to be tossed out into the yard to keep from burning the house down."[14]

Clarence Todd, also of Tulia, recounted another wasteful incident. "The hogs got out once and spilled a five-gallon can of kerosene I had setting next to the smokehouse and left me without fuel for lights," he recalled. "In those early days, I had to go to town by wagon, and it sure was a hard day's trip there and back."[15]

But then the world changed for Swisher County farmers and their counterparts around the state. Suddenly, schoolchildren were doing homework under bright, clean electric lights. Electric pumps brought them indoor running water and made outhouses and water wells obsolete. Women and children no longer had to haul water to the house for cooking, cleaning, and bathing. (In her history of Waco, Patricia Ward Wallace repeated a common joke from the period, a joke that electricity made blessedly outdated: "Yes, we had running water, because I grabbed those two buckets up and ran the two hundred yards to the house with them.")[16]

In a tiny hamlet called Douglassville, a few houses and places of business beneath the towering pines of deep East Texas, J. C. "Po-Boy" Morriss ran an auction barn, the largest in the Ark-La-Tex area. Morriss auctioned off cattle, horses, mules, junk, whatever folks in the area wanted to get rid of (including possums and, a couple of times, elephant tusks). He regularly auctioned off 750 cattle in one day. In later years, most of the mules, ponies, and horses went to California as riding stock or to an East Texas sheriff's posse. In the late 1950s, livestock and wagons used in John Wayne's *Alamo*, filmed near Brackettville in far West Texas, were bought through Po-Boy's auction ring.

Morriss, who grew up in a family of thirteen kids, plastered his slogan on the side of his auction barn: "Bring 'em to me. I am the poest and need the moest. Po-Boy." His wife, Mabel, helped her husband with the auction barn when she wasn't busy at the Dew Drop Inn, her café across the road from her husband's establishment. She would be the reason light and power came to Douglassville. She grew up on a nearby farm and moved to Douglassville when she married Po-Boy in 1915. The newlyweds moved into the oldest house in town.

David Frost, who in years to come would serve for many years on the board of directors of Douglassville-based Bowie-Cass EC, knew them both. When he was in grade school, "Po-Boy would come get all us boys to chase cows. He'd pay you a dollar a day, whether you worked two hours or ten. I loved it. It got me out of school." Frost recalled that the Morrisses were energetic and deeply involved in their small community. "She took ahold on everything; he did too," he said. "He was the Methodist church song leader; she was a Sunday school teacher. They were doers."[17]

"I got busy right away in the church community meeting all the people," Mrs. Morriss told Robert Kerr of the *Texarkana Gazette* in 1982 (reprinted in *Texas Co-op Power* in October 1983). "I knew I was located permanently, so I wanted to make the town as liveable as I could." Her first project was to tackle the dismal lack of general health care available in rural East Texas. "You just can't imagine how things were back in the '20s, even before the bad crash in '29," she said. In 1923, she asked the Cass County Commissioners Court to finance a clinic. The result was the county's first public health unit.[18]

Next, Mrs. Morriss took it upon herself to bring electricity to Douglassville and the surrounding countryside, still dimly lit by coal oil and various battery-powered systems. Like farmers, ranchers, and businesspeople in the Piney Woods, the Morrisses got tired of having to cope with their primitive system. On days when auction sales went past sundown, the Morrisses would keep the bidding going by the light of parked cars, a Delco generating plant, or gasoline lanterns. One night in 1935, every means of illumination failed.[19]

That failure sparked the light of inspiration in the mind of Mabel Morriss. Reading the latest issue of the *Atlanta Citizens-Journal*—published in nearby Atlanta, Texas, not Georgia—on a May evening in 1935, the busy, community-minded woman came across a story about the newly established REA. Her eyes lit up when she read about its offer to finance electricity for rural Americans if they could qualify. She took it upon herself to bring electricity to Cass County. In Frost's words, "She became the ramrod for the co-op."[20]

On October 25, 1935, she launched a letter-writing campaign, targeting local elected officials, Congress, and the REA. She mailed her first letter requesting information about rural electrification to the Works Progress Administration chief in nearby Marshall, who referred her to the REA in Washington, DC. For nearly two years she corresponded—politely but insistently—with REA officials. When she first asked the REA for assistance to bring power to the Douglassville, Marietta, O'Farrell, and Cornett communities, officials told her the project would never pay out and that the area should join with nearby Panola County in seeking aid. "That's how much they knew about the area they were turning down," Mabel Morriss would say, noting with understandable impatience that Cass and Panola Counties were separated by Marion County. Still, she followed up on every lead, many of them blind alleys.[21]

"As you can imagine, with many other rural areas competing throughout the United States, it was unthinkable that this small town in northeast Texas would ever have such a facility," Tommy Cranberry, Douglassville's mayor in 1983, told Kerr. "However, these other areas were actually at a disadvantage, because they didn't have Miss Mabel."[22]

In addition to the boxes and boxes of letters she wrote, the determined East Texan drove her Model T Ford along red-dirt roads through the tall pines, calling on farm families who were trying to make a meager living on plots of land they had hacked out of the forest. "I'd pack my lunch and be gone all day," she recalled years later. "Once I got lost and came out of the woods near dark. I drove on home—and met Po-Boy coming on foot to hunt for me." Morriss used every argument she could dream up to persuade her neighbors to join the effort to organize an electric co-op. As in the Hill Country, the local county agricultural agent would come along behind her and raise doubts about the expense of the project. He would tell the people she had just talked to, "If you listen to her, she'll have you thinking you can plow with electricity!"[23]

Morriss surmised that the REA was getting some of its negative information about Cass County and environs from the Department of Agriculture, since soon after the project was approved, the county agent resigned and moved away. Whatever the agent's motives, there was some justification for his reservations. By the 1930s, cotton farming had fallen off precipitously after decades of planting the same crop in the same fields every year had worn out the soil. Population declined. Hundreds of East Texas farmers left the land.[24]

Morriss faced quite a challenge, Frost recalled. "Cass County is full of trees and hills. It's expensive to set lines. And farmers didn't have many appliances," he said.[25] She wrote to Texas A&M College for assistance and was told the college could give no advice until the loan had been approved. During this period, Bowie County, which adjoins Cass to the west, requested to be included in the first loan, should it be approved.[26]

In February 1936, the REA's Morris L. Cooke again refused to fund the project. Mabel Morriss refused to quit, even though the application continued to be turned down by REA administrator John Carmody. He deemed it "financially unfeasible."

Why was it "financially unfeasible"? That's the question Morriss and her fellow East Texans repeatedly asked. From Washington, the reply was that the area was "sub-marginal" and that the people had no reliable year-round income. The REA claimed its information came from the Department of Agriculture. "The people in Washington thought we were just the poorest people in the world, and they said we couldn't ever afford to pay back a loan like that," Mrs. Morriss recalled.[27]

The arrogance of the bureaucrats in Washington fueled her determination. She went to work lining up financial support in the community, writing to Washington that if her fellow East Texans said they were going to do something, that's exactly what they would do. "Sometimes I would get a hot letter back," she told Kerr, "so I would get on the phone to [US representative] Sam Rayburn or Morris Sheppard or Wright Patman. I said, 'You go tell them what kind of people we've got down here!'" Despite the challenges, Cass County's prospects began to brighten when the REA sent a representative to visit the area, although the initial news was alarming. "You haven't done anything!" the REA man told Morriss. "If you don't get some information in within a week, your project is dead!"[28]

Until then, Morriss had not been told precisely what the REA required. With something tangible to go on, she collected statements from every banker in the project area affirming that the farmers of Bowie, Cass, and Morris Counties had bank accounts as well as year-round income from milk, corn, cotton, and other produce. And yes, they could pay for their electricity.

Traveling every road and rutted lane in Bowie and Cass Counties, she and Po-Boy signed up the required three people per mile. Because at least two-thirds of the farmers in East Texas had annual incomes of $1,000 or less, most found it difficult to come up with five dollars in cash. Somehow, Mabel Morriss persuaded them to make the investment, even if it meant the wife dipping into the butter-and-egg money she had been saving for the kids' school clothes.

W. T. Hammock was one of the early applicants during that crucial week when the co-op's future hung in the balance. Hammock pledged to use six lights, an iron, and a radio.

Within the week, REA officials in Washington were mulling over the bankers' statements and the Bowie-Cass County applications. Carmody approved the project in August 1937, but he wasn't altogether persuaded.

"I know this thing will never pay out, but this is the only way to get that woman off our necks!" he grumbled.

"Mabel connected politically," Frost recalled, "primarily through Wright Patman. She went to D.C. regularly. When they finally signed the agreement, they said, 'Go ahead and sign it; otherwise, she's going to live with us.'"[29]

Not long afterward, the REA informed the indefatigable East Texan that the newly organized Bowie-Cass EC would be receiving a $174,000 loan to build its initial line. The new organization set up shop in a duplex Mabel Morriss owned near the Dew Drop Inn. In February 1939, the first lines were energized. The first billing went to 241 members for a total of 5,717 kilowatt-hours carried over 124 miles of line. The total billing was $553.01. In December 1939, 315.5 miles of line were built and energized, and billing had grown to $2,265.01. By 1945, 636 miles were in operation and the membership was 2,496. The bill was $9,261.72 for 153,406 kilowatt-hours.[30]

Mrs. Morriss was named to the co-op's first board of directors and served as secretary-treasurer, a job she held from 1939 until her retirement in 1978. At the same time, she kept the books for her husband's thriving auction barn, raised two children, tended an orchard and a 150-bush rose garden, and helped out at the general store her husband and father-in-law ran. She also taught Sunday school classes for sixty years. "I wanted to help us get lights for the sales [at the auction barn]," she told *Texas Co-op Power* in 1961. "I expected to help everyone else at the same time, but I didn't expect the co-op to grow this big, but I'm very glad we did it." In addition to all the work Mabel Morriss did for Bowie-Cass EC, she served either as city council member or Douglassville mayor from 1958 to 1981. When she announced she was retiring after nearly a quarter century, her name was dutifully omitted from the ballot. But when the votes were counted, "Miss Mabel," eighty-five at the time, had been elected anyway—by a landslide. She served one more term and then retired again, more insistently.[31]

"I'LL GET IT FOR YOU"

When the early Spanish explorers wandered across the Texas plains in the 1500s, they encountered a river that was reddish in color, so they called it Río Colorado (Spanish for "red river.") Although the earliest explorers didn't know it at the time, the silt-laden waters of the Río Colorado flowed from the western border of Texas, crossed the central plateau and the coastal plain, and then, six hundred miles from its source, emptied into the Gulf of Mexico at Matagorda.

The river was hugely important to early Texans. Stephen F. Austin's colony was located between the Colorado and the river the Spaniards christened "Los Brazos de Dios" (The Arms of God). The two fertile river valleys were the site of the first farms in Texas. Along the banks of the Colorado, President Mirabeau B. Lamar hunted buffalo, and he insisted that the capital of Texas should be a settlement on the river known as Waterloo. His fellow Texans agreed, and the new capital city was renamed Austin, in honor of the "Father of Texas."

Early Austinites soon learned what the Indians knew, what pioneers downriver who were trying to farm the fertile soil soon discovered—that is, that the Colorado, although placid, even dry at times, could become a raging torrent. Homes, property, and livestock were swept away; lives were lost. Between 1900 and 1936, floods ravaged the river valley three times, causing $800 million worth of damage, killing a hundred people, and toppling a dam at Austin.

The state had created the Lower Colorado River Authority (LCRA) in 1935. Three years later, two of eventually six dams—Buchanan and Inks—

were ready to tame the unpredictable river. Mansfield Dam would be completed in 1942. Money for the dams had been appropriated for the stated purpose of controlling the river. Left unsaid for the most part was another purpose: providing cheap public electric power to the constituents of young Lyndon Johnson. By not talking about the second purpose, LBJ image maker Ray Lee explained years later, "they could get the power people off their back."[1]

Johnson had run for Congress in 1937 following the death of US representative James "Buck" Buchanan. "Nobody knows him. He's just a 28-year-old boy," Austin mayor Tom Miller pointed out, also noting that the young man came from the smallest county in the ten-county Tenth Congressional District, a district that included Austin. His campaign speeches included a promise to the people of his district that he "would see that they got electric lights." His first meeting in Washington as a new congressman was to commission federal officials to conduct a power survey to determine which cities and individuals would be potential customers.[2]

He insisted the dams had several purposes, "but the most important . . . was electricity to replace the kerosene lamp." Once the power was available, Ray Lee explained, they would set up a rural electric co-op to distribute it. Power would be supplied to the co-ops at rates averaging about 50 percent of the wholesale rate of private utilities before the dams were constructed. Pedernales EC would deliver LCRA-generated power to rural Hill Country residents to the west, while Lower Colorado River EC—headquartered in Giddings, once it was organized—would deliver power to counties east of Austin.[3]

That was the plan, since the two private utility companies serving the region, Texas Power & Light and Central Power & Light, had shown time and again that they had no interest in serving rural areas. The two companies fought the very idea of rural co-ops and sued to stop the whole Colorado project while Johnson was trying to tie down funding for it. The young congressman and FDR protégé became their bête noire. They hated him almost as much as they hated Sam Rayburn.[4]

The *New York Times* labeled the dams and the power they would produce "the little TVAs of the Southwest." The private companies had another word for the massive project: socialism (or communism).[5]

LBJ biographer Ronnie Dugger explained the issue at hand: "Were private companies to be allowed to build the dams and pile up profits from the power generated from the publicly-owned falling water? Or would the government own the dams and sell the power cheaply to its citizens?" Johnson, avid apostle of the New Deal, was the personification of public power. "They hated me for these dams," he told Dugger. "The power companies gave me hell. They called me a Communist."[6]

He wasn't a Communist, but he was an heir to the radical tradition of late-nineteenth-century populism, a tradition that would inspire electric co-ops, as well. Lyndon Johnson—and certainly his father, Sam Johnson—would have been aware of the People's Party, the Farmers' Alliance, the Texas Farmers' Alliance, the Grange, and cooperative efforts modeled after the Rochdale Pioneers.[7]

"When Lyndon Johnson was a boy in Central Texas, dusty men still trooped past Sam Johnson's white frame house in Johnson City, cursing bankers and muttering about drought," Tom Wicker wrote in the *New York Times* during the early months of the Johnson presidency. "Years later, when he told a group of Texans in New Deal days that private interests were not going to keep rural electrification out of the state 'even if we have to go out and hit them over the head with beer bottles,' he was talking the authentic language of western radicalism, that radicalism that once urged its followers 'to raise less corn and more hell.'"[8] Despite the opprobrium, Johnson went ahead, and the dams got built (without any reported incidents of beer-bottle bashing).

Getting power to rural Texans living along the Colorado turned out to be as difficult for Johnson, despite his Washington connections, as it was for other co-op pioneers. He encountered the same sort of skepticism. At an organizational meeting of the Pedernales EC held at the courthouse in Johnson City, the Gillespie County agricultural agent lectured native son Johnson, telling him he wasn't doing farmers any favors by trying to bring power to the isolated Hill Country. As Caro recounts, the agent, a man named Henry Grote, told Johnson there was no way that farmers would be able to pay their electric bills. They would sink into a veritable quicksand of debt. They would lose their farms. (LBJ must have held his famously volatile temper.) Sixty people in attendance listened to Grote's dire warnings.[9]

Others at the meeting weren't particularly interested because they thought electricity meant only one thing: a naked light bulb dangling on a wire from the parlor ceiling. Standing in front of the county judge's bench, Johnson explained that electricity had countless uses. It could light up the house, of course, but it also could run farm equipment, he told the gathering of farmers. Others simply couldn't afford the five-dollar sign-up fee.

Still others were afraid of the wires. "The idea of electricity—so unknown to them—terrified them," Caro writes. "It was the same stuff as lightning; it sounded dangerous—what would happen to a child who put his hand on a wire? And what about their cows—their precious, irreplaceable few cows that represented so much of their total assets? "And they were afraid of the papers," Caro continues, "the papers that they were being asked to sign. The people of the Hill Country were leery of lawyers: lawyers meant mortgages and foreclosures. Legal documents—documents they did not understand—turned them skittish: Who knew what hidden traps lay within them?"[10]

Skepticism, apprehension, and anxiety weren't confined to the Hill Country. In southeast Texas in 1939, few farmers in Jackson County had ever heard of the federal government making funds available to help farmers help themselves, and few expected anything to change. Area businessmen were convinced that farmers not only didn't need electricity but wouldn't know what to do with it if they got it.[11]

Back in the Hill Country, Burnet County rancher E. "Babe" Smith wanted to be a supporter, but like his Gulf Coast–area counterparts, he had misgivings. In 1936, when he heard that the REA was organizing a co-op in Bell County and that lines would extend to Bartlett, only forty miles from his ranch, he drove over to see whether they could be extended those extra forty miles. No way, he was told. Northeastern Burnet County, where he ranched, was just too thinly populated.[12]

When the Pedernales EC was being organized, his neighbors asked Smith to join the Burnet County delegation that would attend the PEC meeting in Johnson City. He refused. "I said, 'No sense my going.' I been to Bartlett. No hope for me.'" Roy Fry, a politically active Burnet druggist, asked Smith to meet with Johnson. Smith, Fry, and the young congressman got together at Fry's Rexall Drug Store. As Smith recalled, Fry sat on a stool, and he and the congressman sat on packing cases. When a customer came through the front door, Fry would get up to wait on the person.[13]

Johnson reminded the two men that the Buchanan and Inks Dams across the Colorado were almost completed, and with the REA's help, power could finally reach Smith's corner of the county. Smith told Johnson he didn't believe the area would ever get an REA loan. "I'll get it for you," Johnson told him. "I'll go to the REA. I'll go to the president if I have to. But we *will* get the money!"[14]

When county agents and community leaders got together a second time in Johnson City, Smith was in the audience. "Lyndon Johnson had inspired me," he told Caro. "He had made me feel there was a chance."[15]

The people Johnson was trying to reach weren't among the earliest recipients of the REA. Not wanting to take chances on a cooperative failing in the early days, the REA stipulated that its loans would go only to build lines with an average of at least three hookups per mile. One REA representative told PEC organizers in early 1938: "You have too much land and not enough people."[16]

Johnson, whose district stretched from Blanco County in the Hill Country down the Colorado to Washington County southeast of Austin, knew firsthand what it was like to live without electricity. He was born in 1908 near the small Gillespie County community of Stonewall, where the Johnson family farmed and where no one had electricity. When he was five, the family moved to nearby Johnson City, the county seat of adjacent Blanco County. With a population of about three hundred in 1913, Johnson City was larger than Stonewall, but the town was bereft of electricity as well.

The Johnsons weren't farming anymore, but otherwise town life was little different. Mrs. Johnson cooked on a wood-burning stove and washed the family clothes in a tub out back. Kerosene lamps provided dim, flickering light. The family hauled water into the house from a well in the backyard. They used an outhouse. "I remember a few years ago," Johnson liked to say, "when I would have to go out on the back porch, get the coal oil can, take the old potato off the spout and fill the lamps in our house."[17]

Johnson City finally got electricity, after a fashion, in 1927. Johnson (unrelated to the town's founders) by then was a freshman at Southwest Texas State Teachers College in nearby San Marcos, so he wasn't around when Texas Power & Light reluctantly agreed to wire Johnson City and several other Hill Country towns. The company installed a thirty-horsepower diesel generator, which, as Caro writes, provided "only enough voltage for 10-watt bulbs, which were constantly dimming and flickering—and which could not be used at all if an electric appliance (even an electric iron) was also in use. Since the 'power plant' operated only between 'dark to midnight,' a refrigerator was useless." Caro continues, "To most of the residents . . . such problems were academic: so high were TP&L's rates that few families hooked up to its lines. And in any case, the diesel engine was constantly breaking down under the strain placed on it."[18]

In May 1938, representatives from Blanco, Burnet, Gillespie, Hays, and Llano Counties met in Johnson City to form the Pedernales Electric Cooperative. They chose eleven directors—three from Gillespie County and two each from the other counties. They decided Johnson City ought to be headquarters because it was close to the homes of the newly elected president, Hugo Weinheimer of Stonewall, and the secretary-treasurer, Richard Klappenbach, a native of Johnson City. The newly elected board prepared and mailed the loan application to Washington. They waited and wondered whether enough of their neighbors could be persuaded to sign up to make the co-op a reality.

"I'll get it for you," Johnson promised over and over, no matter how skeptical his audience. He traveled the far-flung district with county agents and community leaders. They drove for hours at a time on rutted gravel roads and unpaved lanes to talk to sheep and goat ranchers, dirt farmers, anybody who needed power but might not realize it. Johnson tried every trick of persuasion he could think of. At a meeting in the Kempner schoolhouse with about a hundred farmers in attendance, the congressman played on the emotions of farmwives.

"He talked about his mother, and how he had watched her hauling buckets of water from the river, and rubbing her knuckles off on the scrub board," Caro writes. "Electricity could help them pump their water and wash their clothes, he said. When they got refrigerators, they would no longer have to 'start fresh every morning' with the cooking. 'You'll look younger at forty than your mother,' he told them."

Still, they wouldn't take pen in hand and sign on the dotted line. "What's the matter with these damned people?" Johnson would grumble as the car headed to the next crossroads community. Despite months of effort on the part of county agents and others, nowhere near the number required had signed up. The REA's density rule already had been a stumbling block for the fledgling Pedernales EC.

As Caro tells the story, Johnson had one more card to play. Somehow, he wangled an audience with President Roosevelt. What the brash young congressman said in later years about that audience with the president depended on when and where he was telling the story and to whom. According to one version, the president called REA administrator John Carmody while Johnson was in the Oval Office and told Carmody to go ahead and approve the PEC loan, even though the proposed co-op didn't have the requisite members. "Those folks will catch up to that density problem because they breed pretty fast," Roosevelt may (or may not) have said.[19]

LBJ biographer Dugger offered a more detailed version of Johnson's audience with the president. "I had been told that he liked to see pictures and drawings of ships and maps and things of that kind," he said, "so I had a big picture made of the dam, and a picture of the transmission lines that led from the dam to the big cities." As Johnson described the course of the conversation, he knew that Roosevelt was aware the young Texas New Dealer was about to ask for a favor, so he lingered over Johnson's show-and-tell illustrations. "Great, wonderful, Lyndon!" he exclaimed. "I have never seen better or more marvelous examples of multiple-arch dam construction. It's real ingenuity."[20]

FDR kept talking, and Johnson kept listening, not saying a word.

"Finally, the president ran out of gab," Dugger writes, "fitted a cigarette into his holder, lit it up, and sat back. 'Now, Lyndon, now what in the hell do you want? Just why are you showing me all these pictures?'"

Johnson responded: "Water, water everywhere, not a drop to drink! Power, power everywhere, but not in a home on the banks of these rural rivers."

"What do you mean by that?" an exasperated Roosevelt asked him.

Johnson explained that the people of the Tenth District had supported Roosevelt, but they couldn't get power because of the density rule. He told the president about growing up in the Hill Country, about how he and his brother Sam had toted water from the well and heated it over a wood fire so their mother could wash clothes. He told him about cleaning the smudged kerosene lamps.

"Touched," Dugger writes, "Roosevelt rolled his wheelchair to a phone, telephoned John Carmody . . . and asked him about Johnson's district." Johnson remembered him questioning Carmody about density and then making the remark about Hill Country breeding habits.[21]

Whatever the true story, a telegram arrived at the temporary Johnson City office of the Pedernales EC on September 27, 1938. It brought good news. The REA had granted Pedernales a loan of $1,322,000 (soon to be raised to $1,800,000) to build 1,830 miles of electric lines that would bring electricity to 2,892 Hill Country families.[22]

Ray Lee told Dugger that the struggle to bring power to the people was "a hard, mean, bloody fight, with Texas Power & Light at one end of the district and Central Power & Light at the other."

"When we got our charter and formed our cooperative, outlined where we were going to build our line," Johnson told Dugger, "overnight [the power companies] moved two hundred men into a territory that had never seemed attractive to them before, but they built that line so that it would parallel ours when it was built."

Johnson said he tried to work out an accommodation with the president of one of the power companies, but after several months of fruitless discussion, he finally lost patience. "We've waited many years to get lights in our homes," he told the executive, "and now that we have the chance, you're going to deny us that chance, and I'm not going to try to reason with you any more, because you're so obstinate that I'm convinced it does no good, and so far as I'm concerned you can take a running jump and go straight to hell."[23]

The towns in the co-op service area had to decide whether to stay with the private companies or go with the co-ops. Johnson crusaded for the co-ops as if he were running a reelection campaign. The REA won twenty-five of twenty-six municipal elections over TP&L. Johnson took credit, boasting that he lost only Columbus.[24]

Less than a decade later, Johnson City was also boasting, as Liz Carpenter (a future First Lady's press secretary) noted in a *Corpus Christi Caller-Times* article. "Night motorists in Central Texas are always surprised right outside the city limits of Johnson City to see a large sign ablaze with hundreds of light globes proclaiming proudly, 'Johnson City, Home of the World's Largest REA Cooperative.'" More than three-quarters of a century later, it remained the largest.[25]

In August 1982, *Texas Co-op Power* columnist and former state senator Walter Richter noted: "President Roosevelt called it right, for today Pedernales Electric Cooperative, smack dab in the heart of LBJ country, serves more than 50,000 meters, more than any other cooperative in Texas. However, not all of this growth came from quick breeding; much resulted from folks moving into an area made both attractive and profitable because of this government loan program."[26]

Although Johnson is most often associated with Pedernales EC, serving the Hill Country where he grew up, he also felt an obligation to constituents who lived along the lower Colorado, southeast of Austin. Attorney Sim Gideon, a longtime Johnson associate, recalled in a 1968 oral history interview that Johnson felt it was his obligation, and he wanted to bring the benefits of electricity to his constituents at a reasonable price. "He felt like that it had to be brought out to the farms and ranches; otherwise the farms and ranches were going to disappear," Gideon recalled. "People weren't going to live on them unless they could have the benefits of electricity."[27]

Initially, the co-op along the Colorado east of Austin was called the Lower Colorado Electric Cooperative, but frequent confusion with the Lower Colorado River Authority (LCRA) prompted a name change to Bluebonnet Electric Cooperative in 1961.

Looking for a ramrod for the lower Colorado co-op, someone with the energy and enthusiasm of Mabel Morriss in East Texas, the congressman settled on Carl Angus McEachern, a farmer and rancher in rural Travis County, near the Webberville community. Like Morriss and other co-op pioneers around the state, McEachern was busy, but never too busy to serve his community. Growing cotton and raising cattle on an 1,800-acre farm, he also served on several bank boards and was a deacon at Hyde Park Baptist Church in Austin. His co-op influence over nearly four decades would rival that of the Hill Country congressman who had sought him out. Like Morriss, he was a natural-born leader.[28]

In his history of Bluebonnet Electric Cooperative, published on the eightieth anniversary of the organization, Ed Crowell recounted an interview McEachern gave in 1979. McEachern, known as C.A., recalled that a man showed up at his house one morning with a message: "Lyndon Johnson, your congressman, wants you to come into town. He wants to organize an electric co-op."

On that particular morning, McEachern wasn't interested in driving into town (Austin), or organizing an electric co-op. "I said, 'Listen fellow, I'm a farmer and in agriculture; I don't know anything about electricity.' And the man laughed and said, 'Well you don't have to know anything about electricity. He says he wants you as one of the directors, and if you could be at that [organizing] meeting, why he'd appreciate it.'"

McEachern drove into Austin and met with Johnson and other future board members. The meeting started at 2:00 and didn't last long. Mostly, McEachern recalled, they listened to Johnson lay out what he wanted to do and how he intended to do it.

Later that day, McEachern got a call from Gideon, who would become general counsel and later LCRA's general manager. Gideon invited him to dinner that night at his house and also asked him to go into Austin and pick up Johnson at the Driskill Hotel. McEachern accepted the invitation.

"Going out there," McEachern recalled, "[Johnson] told me, 'This co-op is going to need $100,000 to start on. . . . I'm going back to Washington, and in a day or two I'll call you.'" McEachern got a call a couple of days later, just as Johnson had promised. "'I got it,' the congressman told him. 'You fellows just go right ahead now and start your business and start working on it and I'll see you later.'"

McEachern put up his own money to hire a few employees to build the first extensions from already existing lines. On August 1, 1939, Johnson sent a telegram to the publisher of the *Austin American-Statesman* announcing that the REA had agreed to finance 825 miles of lines to serve 2,125 co-op members east of Austin in Lee, Washington, Travis, Bastrop, Fayette, Austin, Williamson, Guadalupe, Hays, and Caldwell Counties. (Later, parts of Burleson, Colorado, Gonzales, and Milam Counties were added to the Bluebonnet service area.) The loan was $665,000. Two days later, the newly appointed board of directors for the Lower Colorado River EC held its first official meeting at the Littlefield Building in downtown Austin. The board elected McEachern president.

The hardworking farmer-rancher would be an indispensable co-op member, serving as board president for nearly four decades, Crowell noted in his Bluebonnet EC history. "He helped steer the organization through its initial growing pains and then a near halt in new hookups during World War II. He approved line construction and operating agreements with the LCRA and then a total separation from the agency other than as a power source."

He had a hand in every decision from the beginning, including where the co-op headquarters ought to be. McEachern said in the 1979 interview that the board realized after a few meetings in Austin that the co-op needed a headquarters building outside the city. Several towns were interested, and their representatives approached McEachern with offers. "I said, 'Well now listen, you go talk to Mr. Johnson,'" he recalled. "I said he was ramrodding this part of it, and Giddings was almost the center at that time [of the original service area], so the office was finally set up in Giddings."

McEachern also relied on his ties to Johnson, who used his connections to the National Youth Administration to get the headquarters built. The congressman enlisted NYA labor and relied on NYA workshops to construct the furniture. The congressman and then US senator stayed in touch with the co-op through the years, writing letters of welcome to new members and showing up at annual meetings, particularly in election years. In a 1949 speech in El Paso, the newly elected US senator Johnson expressed his admiration for rural electrification. He didn't have to tell his listeners how hard he had worked to make the REA a success. "REA stands today as democracy's most successful experiment in faith and hope—and there is no charity involved," he said, noting that co-ops already were repaying their loans. "In little more than a decade, we have turned on the lights in more than 3 million American farm homes—and it hasn't cost the taxpayer a cent."[29]

Turning on the lights was just part of the story, as Jo Nell Schulze recalled in 2019. She grew up on her grandparents' farm outside Lockhart. Before electricity, her grandmother or grandfather would carry coal-oil lamps from room to room. Years later, she still had the base of one of those lamps, as well as the grandparents' old wooden ice box. "Granddaddy would go to town on Saturday," Schulze recalled, "and he'd take a tote sack with him, and he would go to the ice plant and get a big block of ice and put it in that ice box. And, hot dog, we were going to have iced tea then! That block lasted about three days. My grandmother kept her milk and eggs in the top shelf."[30]

Mildred Kramer Richter, born to a German immigrant family in 1921, still had vivid memories nearly a century later of the day the power arrived at the family farm in Washington County near the Prairie Hill community. "Here we saw these people come into our yard," she recalled in a 2019 *Texas Co-op Power* article, "and we thought, 'What in the world are they doing?' And all at once we saw them put poles up, and they were walking on the poles with cleats, and they had these little green cups made out of glass that they connected way on top." Once Bluebonnet EC got the Richter farmhouse hooked up—Bluebonnet at the time was still the Lower Colorado River Electric Cooperative—the family of ten kids embraced the electric age. "My dad was this way: If he could find anything that worked with electricity, he would buy it," Richter recalled. "A washing machine and a stove and a refrigerator." Her father bought the appliance at either Sears or Montgomery Ward in Brenham.[31]

Many of the early rural co-ops relied on "ramrods" like McEachern at Bluebonnet and Morriss at Bowie-Cass, individuals who believed in the mission, who enjoyed the trust of their neighbors, and who had the energy to make it happen. The ramrod for Hill County EC (later called Hilco EC) was Earl D. H. Farrow, who, as a teenager, tried to persuade TP&L to build a line to the Farrow family farm. The Farrows were sharecroppers. Like their rural neighbors, they still relied on candles and a kerosene lamp to light the house, still had to cut wood to heat the house in winter and feed the cookstove year-round. Every endless chore they had to do on the farm they had to do by hand because they had no electricity.

On a hot summer day in 1911, the sixteen-year-old Hill County farm boy was picking cotton on land the Farrow family sharecropped outside Itasca. Perhaps he stood up to wipe his brow and stretch his aching back, but however it happened, he noticed a group of workers putting up power lines about a quarter mile down the road. It was the local utility company, Texas Power & Light, stringing wire to homes in Itasca.[32]

Young Farrow left his cotton sack at a turnrow and wandered over to the lineworkers. "We need power too," he told the men. "Can you build a line to our house?"

"Can't do it, son," he was told. "TP&L don't come out this far. Cost way too much money to put up miles of poles and string wire just to bring power to the few folks who live out here."

Maybe the teenager argued; maybe he didn't. Maybe he trudged back to the cotton field that fateful day, already dreaming of a way to get electricity to his folks and their neighbors. Maybe that's where it started for E. D. H. Farrow. It would happen, but it would take twenty-five years before his dream came true.

At age twenty-two, he escaped the endless rows of cotton and corn—for a while, at least—by joining the navy and serving his country for two years during the Great War. He completed his tour of duty in 1919; married Pauline Lofton, a young Red Cross volunteer he had met during the war; and on Christmas morning 1919 took her home to Texas.

The young couple tried to make a go of it on the family farm, with Earl doing odd jobs to supplement their meager income. The Farrow farm was still powerless, even though they could see a TP&L substation from a corner of their cotton field. In 1921, Earl Farrow was working a temporary job paving Itasca streets when a local business owner offered him a job as a salesclerk in his furniture store. Earl took the job, and with the stability of a steady paycheck the couple managed to move into an apartment in town (with electricity). In 1926, he became the town's undertaker, selling furniture on the first floor and embalming people upstairs. Two years later, he was elected mayor of Itasca. He was thirty-three.

"He served eight years as mayor of Itasca," Brian K. Moreland writes in his history of Hilco EC, "doing his part to keep the town from blowing away with the dust storms like so many small towns during the Depression. Little did Mr. Farrow know that his career was about to change radically in the summer of 1936, when a man from the REA showed up at his door."

The REA man from Washington approached Farrow not only because he was mayor, but also, no doubt, because he was a trusted member of the community. Farrow agreed to help recruit farmers, and after he had signed up quite a few, agreed to go to work as the organizing manager for the proposed co-op. With plans for a headquarters in nearby Hillsboro, the county seat, the co-op would cover five counties.

Farrow faced a dilemma, though. If he accepted the job, he would receive $150,000 in seed money—but he would have to quit his furniture job and work full time for the REA without getting paid for the first six months. His only source of income would be embalming, part time but steady.

As Moreland tells the story, Pauline Farrow assured her husband that he should do whatever he wanted to do. His friend the local grocer assured him that he could run a tab until he started getting paid. His banker friend told him, "If you need money, we'll lend you some." And the local auto mechanic told him, "I've got a truck down here you can use, and I'll give you gas."

Farrow took the job, on condition that the co-op remain in Itasca. Request granted, he named the new venture Hill County Electric Cooperative, hired a secretary and a man to drive around the territory and persuade people to sign up, and appointed a board of directors. The first REA check arrived toward the end of the year.

Earl Farrow's son Gene told Moreland that on that happy day his father took him downtown to the local café, where he told the waitress he'd take care of the bills for everybody in the place. And then he asked her whether she could cash his check, his $150,000 REA check.

"That was a glorious day for Earl Farrow," his son recalled. After celebrating with his neighbors, he deposited the check in the bank and repaid everyone he owed. He would spend the rest of his life working for the co-op. Earl D. H. Farrow retired in November 1968. He died a month later.[33]

In 1939, REA administrator John Carmody received a letter from a West Texas rancher that captured the growing enthusiasm and momentum at decade's end. "I have been in various parts of Texas where there have been oil booms," the rancher wrote, "where enthusiasm was at fever heat, here it cost two dollars a plate for ham and eggs, and here one was glad to pay five dollars to spend one night in a barber chair. But I have never seen such enthusiasm as is now shown in Deaf Smith and Castro counties over the coming of REA electricity."[34]

Three Texas co-ops were established in 1936 and seventeen in 1937 (including Hilco). Pedernales EC was one of twenty-seven organized in 1938. Thirteen systems were organized in 1939 and a number of others in the early 1940s. Rural life would never be the same.

"This was a threshold, a beginning," Marquis Childs wrote in *The Farmer Takes a Hand*. The farmer—and the farmer's wife in Cass County and elsewhere—had indeed taken a hand, had risen up and had done, for himself and herself, what had to be done to bring light and power to the countryside. The powerful investor-owned utilities had done nothing for rural Americans over the years. Now that co-ops were up and running, the investor-owned utilities prepared for war.[35]

"I TOLD 'EM TO GO TO HELL"

In the second decade of the twenty-first century, about the only business left in Zabcikville, a tiny Czech immigrant community on Possum Creek south of Belton, is Green's Sausage House. A modest building, the combination café and meat market is surrounded in spring by fields of corn as high as a horse's eye, if not an elephant's. At lunchtime its gravel parking lot is packed with cars and trucks. Green's attracts customers from nearby Temple, Belton, and Killeen, as well as points farther afield.

Frank Zabcik (pronounced ZAB-chik) started the business in the 1930s and then sold it to Jerome and Della Green in 1946. The partners first operated a café and grocery store, as well as a granary business. Today, Green's Sausage House, known for its sausage, kraut burgers, and beer-battered onion rings, is owned by Charles and Marvin Green, sons of the original owners. The Greens are known for their twenty-two types of sausage, based on old Czech recipes, but Green's Sausage House these days is a combination café, bakery, meat market, and twenty-four-hour drop-off deer processing service.

The Czech immigrants to eastern Bell County, most of them farmers, came principally from rural regions of Moravia, where the soil was thin, poor, and depleted. On the blackland prairie of Central Texas, they found some of the most fertile soil in the state, some of it ten to fifteen feet deep.[1]

They also rooted themselves in freedom's soil. No longer would they be subject to a feudal system that required them to pay manorial dues to the nobility, the state, and the church. The most despised obligation was *robota*, an ancient tradition binding the peasant to work free for the lord of the manor for a specified number of days annually.

Arriving in Texas in 1876, the Czechs established Church of the Brethren congregations, expressing their allegiance to a spiritual heritage dating back to the fifteenth-century pre-Reformation teachings of Jan Hus. (The Czech theologian and reformer, born in 1369, was burned at the stake for his beliefs in 1415.) Their well-preserved churches, usually graceful white-frame buildings with a steeple visible for miles around, continue to prosper, even if Ocker, Cyclone, Seaton, Zabcikville, and other small settlements withered away.[2]

From the beginning, the Czechs took pride in their hard work, as evidenced by a common saying among immigrant farmers: "My friend, do you think that feathers grow on me? I do not own a mint for the making of money; I did not find dumplings here on trees. Neither did I find those brooks full of butter and honey about which I had heard in the old country."[3]

The Zabciks were prominent from the beginning. In the early 1900s, Jan Zabcik built the first gin in the community, steam powered. In the early days of the general store that became Green's Sausage House, the owner brought "medicine shows" to Zabcikville. In the mid-1940s, Frank Zabcik built a baseball field and helped organize a team that participated in the Cen-Tex League.[4]

On a rainy summer day in 1936, Adolf Zeistman, A. J. Schneider, and Bill Kelm were sitting around Frank Zabcik's general store talking about crops and livestock and the weather as they waited for the rain to let up. Zabcik raised a vexing issue: "How in the world can we get electricity?"

Kelm mentioned that he had heard about some sort of new government program to help farmers get electricity. "In fact," he said, "I've got a little folder at home that tells about it."

The next day Kelm dropped by the store again, brochure in hand. He and Zabcik studied it and then wrote to Washington, requesting additional information about the REA program they were reading about. Not long afterward, an REA representative showed up at the store. He explained the program in detail to interested farmers.

"The farmers were enthusiastic. They talked it up and really got the ball rolling," Zabcik told *Texas Co-op Power* more than two decades later. "The more we planned and talked about REA, the more Texas Power and Light bombarded us to do business with them. Before we knew it, we practically had a shooting scrape going."[5]

Zabcik and his neighbors, desperate for power for decades, didn't realize it initially, but they were being subjected to a tactic the private power companies would employ across the country in an effort to cripple or destroy the co-ops before they could get established. Their devious overtures to potential co-op members would come to be called "spite lines."[6]

In his book *The Farmer Takes a Hand*, Marquis Childs explained how the tactic worked. Farmers in a relatively prosperous community (like the Czech settlements in Central Texas) band together to establish a co-op because private utility companies have ignored them for years. They send their plans for the project to Washington for approval for, say, twenty-six miles of line to serve perhaps a hundred members. The local power company, hearing the co-op is about to be approved, suddenly decides to drive a seven-mile power line through the middle of the proposed co-op service area, picking up thirty-five or so of the co-op's best customers. "Without these 35 customers, the project is not practicable," Childs points out. "Consequently REA is forced to refuse the loan."

Without the co-op, the remaining farm households come to the private company, requesting to be hooked up too. Sorry, they are told. It's just not profitable to serve any others beyond those already on the close-knit seven-mile line. "If this had happened in a few isolated instances, farm leaders would not have come to suspect that this was a deliberate strategy of the power companies," Childs writes. "But more than 200 co-ops in 40 states were faced with the tactic of the spite line. Eight of these cooperatives were completely wiped out. Others suffered long and hampering delays."[7]

The National Rural Electric Cooperative Association (NRECA) labeled the tactic "pirating." An NRECA survey in the spring of 1953 found pirating being used around the country. "The most publicized fight was in Newport, Ark., where the giant Arkansas Power & Light Company used unlimited funds and mobilized its personnel from all over the state to take from one small rural electric cooperative a large section of consumer members that they had developed where there had been only cow pastures before," *Texas Co-op Power* reported.[8]

Quoting *Time* magazine, *TCP* pointed an accusatory finger at Arkansas Power & Light CEO Ham Moses, whose success at "stifling a tiny adversary with his wealth and facilities" was spawning imitators around the country. The NRECA survey found that rural electric systems in fifteen states were under siege, mostly from private power companies, with municipals not far behind.

"The power companies generally are involved in taking the more desirable loads, such as areas of high density, mines and machine shops," *TCP* reported. "Contrary to the profit power propaganda that only public power builds duplicating lines, in almost every case the profit power company doing the pirating builds parallel or duplicating lines themselves. Losses to the co-ops create difficulties for them in striving to provide area coverage. When raided, the rural system faces necessity of redesign of the system, loss of revenue and even lack of compensation."[9]

TP&L was an energetic spite-line builder. Crews worked all night erecting lines neck and neck with completed REA lines. In some places their lines were as close as two feet from each other. Often lists of the most lucrative rural areas were covertly passed to the private companies, so that, like cowbirds taking over a nest, they could move in ahead of the co-ops. Spite lines were often constructed out from urban areas in a pinwheel pattern along main roads in the lucrative outlying areas, leaving to the co-ops the poor, isolated houses between the spokes. The result was that co-op formation was difficult or impossible.[10]

Private utility salespeople, like the indefatigable Fuller Brush and Bible salespeople, went door to door selling power. They warned potential co-op members that government-financed lines "were all a dream and would never be built." They also warned that if they were built, the government would be able to seize their farms if the lines proved untenable, in exchange for the public investment. TP&L also told farmers that all of their neighbors had "signed for power company service" when in fact "few or none" had. Their efforts convinced some prospective members to cancel their contracts with their cooperatives.[11]

Pressure directed toward farmers around Zabcikville was a textbook case of the spite-line, or pirating, tactic. TP&L would not relent, Frank Zabcik recalled. "They offered to wire anyone's house free. They even offered to install a $1,200 meat counter in my store. I told them, 'No.' I didn't want their electricity. I had done a lot of work for the REA, and we had a good cooperative started."[12]

A. A. Winkelman, one of the founding directors along with Zabcik, got the same power-company treatment. A line went in along the road in front of the Winkelman farmhouse, and a power-company representative dropped by with inducements: abandon the co-op, he told Winkelman, and we'll wire the farm for free, throw in a new refrigerator, and sell the

Winkelman family any additional electric appliances at cost. Apparently, Winkelman was polite. He told the company representative that his neighbors, whose houses were farther off the road than his, placed their faith in him. They were trusting him, A. A. Winkelman, to get electricity to all of them. He had no intention of letting them down. He also reminded the "claim jumper"—*Texas Co-op Power*'s disparaging label for the would-be seducers from the private power company—that he was well aware that the man would not be at his door if all the farmers in the area had not organized their co-op to get service to everyone.[13]

TP&L managed to win a few converts. The company's inducements included offering electricity immediately to any person who signed up and also hiring local men—during the Depression—to string wire. Occasionally, the power company put up poles and lines before getting easements from landowners. Things got nasty. As soon as TP&L got a pole in the ground, someone would come along and chop it down. The two sides went to court. The newly established Belfalls Electric Cooperative prevailed.[14]

Still, the investor-owned utilities persisted across the state. In September 1944, in the third edition of *Texas Cooperative Electric Power* (forerunner to *Texas Co-op Power*), editor George W. Haggard reported that he had visited a couple of annual meetings and had enjoyed hearing what cooperative electricity had meant for them. He also heard stories about what he called "Power Trust skullduggery." Farmers told him how they had begged the private utilities for years to build a line to their farm and how they were inevitably ignored—until the REA came along. Suddenly, the power companies were promising light and electricity overnight. "I told 'em to go to hell," one farmer told Haggard. "I'd been after 'em for ten years to give me lights, and now that I had a chance to own a share in the co-op that would serve me, I really enjoyed telling the power company where to go."[15]

The first meeting of the Belfalls Electric Cooperative—named for Bell and Falls Counties—was held on June 16, 1936. Frank Zabcik was elected president of the five-member board, each of whom put up five dollars. On September 14, 1936, the co-op procured an REA loan for $452,000 to build three hundred miles of line in the Zabcikville-Westphalia area. The line served the original eight hundred members.

On July 1, 2007, members of the Belfalls EC and the McLennan County EC voted to consolidate operations, resulting in a new cooperative called

Heart of Texas Electric Cooperative, based in McGregor, west of Waco. Belfalls no longer exists, but descendants of those Czech farmers who were its founders are still co-op members, in spite of a certain power company's pressure.

"WELL, WE JUST TURNED ON THE LIGHT AND SAT AND LOOKED AT EACH OTHER"

For fifty years the power companies had told the farmer that it was too expensive to wire rural America. Too expensive for rural Americans to use electricity. And even if farmers and ranchers could afford it, electricity would be of negligible value.

Henry Angerstein, a DeWitt County farmer who owned a commercial chicken hatchery, no doubt would have differed, particularly when he was turning the crank and feeding countless corncobs one by one into his mechanical sheller day after day. Like every other farmer he knew, he spent hours shelling enough corn to feed his horses and mules and chickens.

The first hand-operated corn sheller was patented in 1839. To be sure, it was an improvement over scraping shucked ears of corn across a board studded with nails. The newfangled sheller that Angerstein used not only saved a farmer time and energy but also guarded against painful scraped knuckles. Still, it was hard work. On a spring afternoon nearly a century after the invention, Angerstein happened to be shelling corn when his friend C. W. Beck dropped by.

Beck, manager of the newly established DeWitt County Electric Co-op—DeWitt consolidated with Guadalupe Valley EC in 2001—knew that baby chicks had to eat, which meant that Angerstein or one of his employees had to shell corn. He also knew that turning a hand crank on a sheller, feeding the cobs into the machine, and collecting the kernels into a bucket was tedious, time-consuming work. It was particularly hard for a hatchery, where hundreds of voracious little balls of incessantly piping yellow fluff consumed a total for the brood of several bushels a day, on their way to becoming full-grown hens.

"Wouldn't it be great if we could figure out some way an electric motor could do that job?" Beck mused. "That's the sort of thing children could do after school, and it would free you to do other work."

Angerstein paused, perhaps wiped his brow with a bandanna, and glanced down at the bushels of unshucked cobs at his feet while pondering the possibility. The two men talked, and as they talked they came up with a plan. Beck would carry the sheller back to the co-op offices in Cuero, and if he could redesign it for motor use, Angerstein would pay for the cost of the materials. If Beck was unsuccessful, he would buy Angerstein a new sheller, since the old one would probably be junk after his tinkering.

The co-op manager drove back out to the Angerstein farm a few days later. Angerstein noticed that his friend had a smile on his face as he got out of his car. That smile signaled success. He had indeed electrified the corn sheller, using a quarter-horsepower, 1,725-rpm motor with a built-in switch and cord.

The machine worked so well that it not only made Angerstein's life easier from there on out but also served as a prototype for some two hundred other shellers on the DeWitt line. At the time of the article in the July 1944 issue of *Texas Cooperative Electric Power*, Beck reported that he had moved on to building an electric sausage mill. He said that farmers who had used the converted shellers told him that the cost of materials was minimal compared to the time and labor they saved. He said he expected the sausage mill to be just as successful.[1]

Across the state, farmers and ranchers like Angerstein immediately found time- and labor-saving applications for electricity, in the home and out in the barn. In Fannin County, Tommy Johnson in 1934 had been a struggling young tenant farmer growing cotton on a 104-acre spread three miles south of Bonham. Still in his thirties a decade later, Johnson owned the farm, plus an adjoining forty acres, and had more than doubled his annual income. Thanks to electricity.

As a tenant farmer, he had been selling milk from his one cow to a cheese plant in Bonham. Eventually, he was successful enough to give up cotton farming and start dairying full time. Since he didn't have the money to go into the dairy business on a large scale, he changed over gradually, buying another cow when he had the money and planting a little less cotton each year.

When the Fannin County Electric Cooperative extended electricity to his farm in 1937, Johnson was poised to take advantage. With the power to operate a full-fledged dairy business, he could bottle his milk and sell it to Bonham grocery stores and cafés. He bought an electric milking machine, an electric cooler, and other dairying appliances and found his first customer—Sarge's Café in Bonham. Other customers soon followed, and he built up his herd. Milking ten cows a day produced thirty gallons daily. He had all the business he could handle.[2]

Near the Plum community in Fayette County, southeast of Austin, sixty-five-year-old John D. Kovar and his wife, Frances, were farming 227 acres in 1944, a few years after Fayette County EC was formed. The couple deeply missed their grown sons. Kleophas had been killed in action in Germany the previous fall; Francis was confined to a German POW camp.

The Kovars had been farming for forty-two years. Every day of their life on the farm they had drawn well water by hand. Bucket after endless heavy bucket, they drew enough water for eighteen milk cows, several hundred chickens, and assorted other livestock. And then came electricity. The installation of an electric pump transformed their lives. Their old wooden bucket became rustic decoration. If only the boys could have seen the change, they often thought to themselves.[3]

Mrs. Victor Lack of Milsap, in Parker County, may have appreciated electricity more than most. The Tri-County EC member had cooked on a woodstove "forever," she said, even though she had been blind for forty-two years, since age twenty-three.

The morning after the Lack family got their electric range, her husband, Victor, brought in a load of freshly cut wood, just like always. And then it hit him. "Stove's gone," he muttered, his arms full of logs.

"He was so mad," Mrs. Lack laughed, "but he just had to take the wood back out."

"I don't dread cooking now, because I don't have to build a fire or put up with dirt and soot in the house," she told Bud McAnally, the electric-use adviser for Tri-County Electric Cooperative. When an electric range replaced Mrs. Lack's old woodstove, she memorized the feel of the dials and knobs, when they were on, when they were off, and what temperature she needed for what she was cooking. "A slice of her French custard pie testifies to her ability as a good cook," McAnally reported. "Mrs. Lack has recipes read to her, then she 'writes them on my mind' for future use."[4]

M. B. Chandler, a business owner in a Bowie County community called Maud, wasn't too hot on the idea of rural electrification when he saw REA-financed lines going up in 1939. "I was pretty blue about the whole thing," he said, explaining that he was in the ice business in Maud. "I thought the electric refrigerator would run me clear out of the ice business," he said. "But I soon learned I was wrong. People who bought electric refrigerators sold their old ice boxes to people who didn't have any ice box at all, and I kept right on selling my ice."

Chandler also got into the ice cream business, buying $35,000 worth of equipment and hiring about a dozen employees for his new venture. Since his rural customers had started buying electric refrigerators, they didn't have to worry about Chandler's ice cream melting before they got around to taking the first lick. His business was going strong.[5]

Brody Koon, who grew up in the Bonanza community near Sulphur Springs, was a seventeen-year-old partner in the family dairy business when he told *Texas Cooperative Electric Power* in 1944 what happened when power came to Hopkins County. "We started with just lights," he said. "Then, before long, we got an electric milker. We used to allow three hours for doing the milking by hand. The machine cut the time down to just an hour."[6]

Another teenager, Edmund Farrow, made a study of all electric appliances his family used on the Farrow farm near the East Texas town of Linden. Farrow, the sixteen-year-old president of the Cass County 4-H Club Council, noted that farm electricity furnished the lights, pumped the water, churned the milk, toasted the bread, cooked the waffles, mowed the lawn, hatched the eggs, brooded the chicks, kept the cows in their pastures and the hogs in their pens, ground the feed, washed and ironed the clothes, brought news and music over the radio, cleaned the house, cooked the meals, heated the water, and cooled the house in summer.[7]

A 1944 survey by a farm magazine called *Country Gentleman* found that families like the Farrows were typical. Despite the power companies' claims, the magazine estimated that farmers would spend more than a billion dollars on appliances once they got electricity. Plumbing alone—sinks, bathtubs, indoor toilets—was a market worth $350 billion to private enterprise.[8]

In 1951, with rural Texas 89 percent electrified, *Texas Co-op Power* reported that electricity handled six hundred different jobs on the farm. One of them not listed was the element of surprise, as an elderly Floyd County farmer's wife explained, recalling when she and her husband finally got electricity: "Well, we just turned on the light and sat and looked at each other," she said. "It was the first time I had seen Pa after dark in 30 years."[9]

Often it was the little things, like relegating a kerosene lamp to the storeroom, that made a huge difference in a person's life. It made a difference for Milton Collier Phillips, who grew up in Limestone County near a community called Point Enterprise. "The electricity just changed the life around for everybody, and it helped me so much because my eyes were bad," he recalled years later. "And the kerosene lamp, it drew candle flies to it that would really pester you when you tried to get your lessons at night, and it was quite an ordeal. I didn't remember how much of a boost it was when the rural electricity come along until we got it. It just changed everything. I guess it changed the world."[10]

Charles Thatcher of Palacios literally waxed poetic about the coming of electricity to rural Texas. In the February 1946 edition of *Texas Cooperative Electric Power*, he wrote the following lines:

> It hatches my chicks and keeps them warm,
> That's why I'm still living down on the farm.
> It washes my dishes, and also our clothes;
> And waters the garden with a big black hose.
> Farm work now is just like play
> Since we got electricity from REA.[11]

Not every rural Texan was instantly enamored of electric power, as Bryan-based business owner Jordan Lawler discovered in 1925, when he purchased a water-powered gristmill on the Medina River at Castroville. For many years the mill had allowed area farmers, most of them of Alsatian heritage, to mill their grain in Castroville rather than shipping it to San Antonio, forty miles to the east. Two years after purchasing the mill, Lawler converted it to a small hydroelectric plant that provided electricity for the first time to the picturesque Medina County town.

One of Lawler's potential customers turned him away, explaining that she saw no need to alter her lifelong routine of getting up when the sun rose and going to bed when it set, as God intended. Another elderly resident grudgingly allowed Lawler to wire his house, but Lawler had to agree not to bill him until he actually used the service. One night the old fellow got sick, and his wife couldn't find matches to light the lantern. "Dammit," he exclaimed, "turn on the electric light and tell Lawler we are customers."[12]

In 1987, on the fiftieth anniversary of power coming to Marion County in East Texas, seventy-nine-year-old Helen Hicks recalled life before electricity in Lodi, the tiny settlement where she was born and where she had lived all her life. Her father had built the house where she resided in 1906. "I grew up in a farm family of nine brothers and sisters," she told Bill Thompson of Upshur Rural EC. "I can remember only too well all the hot wood fires we used to have to burn to cook with, the old iron flatirons we used, the water we had to heat on a wood stove to wash with and the messy kerosene lamps." She knew all the old ways, she said, adding, "I don't think young people could cope with all those old things we had to do before electricity came. . . . I will never forget those times, but I sure wouldn't want to go back to them."[13]

It would be hard to find a Texas farmer or rancher who longed for the old days, the old ways. As the hard decade of the '30s came to a close, the nation still faced formidable difficulties, but on the farm, at least, electricity was making a dramatic difference. By 1940, 567 cooperatives across the nation were providing electricity to 1.5 million consumers in forty-six states. Rural households with electricity had risen to 25 percent.

In Texas in 1936 a measly 3 percent of farms and ranches were electrified. "The barns on most farms in Sweden were better lighted than the majority of rural homes in this state, and, in fact, throughout rural America," G. W. Haggard wrote in the 1944 inaugural edition of *Texas Cooperative Electric Power*. Eight years later, 30 percent of Texas farms were electrified. The REA at last had brought electricity to places like the Hill Country, where before, in the words of historian Stephen Harrigan, "the paleolithic darkness was relieved only by lanterns and hearth fires." With power's arrival, thousands of rural Texans saw their lives change for the better almost instantaneously. In editor Haggard's words, "No other achievement in history surpasses the record of what American farmers and their families have accomplished during these eight years through cooperative action."[14]

Hyperbole? Perhaps, although Henry Angerstein, Tommy Johnson, John and Frances Kovar, Brody Koon, Edmund Farrow, Milton Collier Phillips, and Helen Hicks were not likely to think so. They would likely have identified with the Smith family and their rural neighbors in the Texas Hill Country.

As Robert Caro tells the story, Brian Smith had persuaded many of his neighbors to pay their five dollars and sign up for the Pedernales Electric Co-op. When more than a year had passed and power had still not arrived, the neighbors decided they had been duped by the government. Smith's daughter Evelyn recalled dropping by to see a friend one day and being told by the friend's parents to leave because they were frustrated with her father. "You and your city ways," they told her. "You can go home, and we don't care to see you again."

"Even the Smiths were beginning to doubt," Evelyn Smith told Caro. It had been so long since the wiring had been installed, they couldn't remember whether the switches were in the on or off position.

"But then one evening in November 1939," Caro writes, "the Smiths were returning from Johnson City, where they had been attending a declamation contest. As they neared the farmhouse, something was different.

"'Oh, my God,' her mother said. 'The house is on fire!'

"But as they got closer, they saw the light wasn't fire. 'No, Mama,' Evelyn said. 'The lights are on.'"

They were on all over the Hill Country. As one resident told Caro years later, "People began to name their kids for Lyndon Johnson."[15]

"GREETINGS"

It was a springtime Sunday morning in Frydek, Texas, in the year 2013. With mass just concluded at St. Mary's Catholic Church, ninety-year-old Johnny W. Konarik stood outside enjoying the sunshine and reminiscing about a letter he had received more than seven decades earlier, shortly after the Japanese bombed Pearl Harbor. The letter to the teenage farm boy was from Uncle Sam; it began with "Greetings." It was his draft notice.

"My daddy didn't want me to go," Konarik recalled, but his country was at war, and he felt called to do his duty. He was one of sixty-five young men from the tiny farming community near Sealy, west of Houston, who fought in World War II. Frydek, served by San Bernard EC, was established in about 1895 by Czech farmers. The earliest settlers in the area arrived in the 1820s as members of Austin's Colony. They patented several "labores" of land along the Brazos just south of San Felipe.

Many of the Frydek sixty-five, as they came to be called—the name is pronounced "FREE-dek"—were in the thick of things. Konarik, a navy man, was on a supply ship that plied the dangerous waters of the Pacific. Several were shot down over Europe, and at least three served time in Nazi POW camps. A few saw action in both Europe and the Pacific.

Remarkably, every one of the sixty-five made it back home alive. The locals, almost all of them Catholic, called it a miracle. They celebrated with an annual spring gathering on the live oak–shaded grounds of the church, near the parish's famous grotto that also commemorated the men's return.[1]

About an hour west of Frydek is Praha, another tiny farming community settled, like Frydek, by Czech immigrants. (Praha is served by Fayette EC.) In the cemetery of St. Mary's Church of the Assumption, one of Fayette County's lovely old "painted churches," a granite memorial erected in 1984 commemorated the nine young Praha men, all farm boys, who went to war. None survived.

All nine lost their lives within about a year of each other, toward the end of the war. Percentage-wise, few cities or towns in America suffered a greater loss. Like Frydek, Praha celebrates those young men, forever young, with a yearly special service.[2]

On the flat and treeless Panhandle plains west of Amarillo, the residents of Hereford had a slightly different WWII experience. Hereford, home since 1937 to Deaf Smith EC, not only sent soldiers and sailors to fight but also received the enemy. The little county seat was the site of the second largest of the seventy-five POW camps built during World War II.[3]

The newly formed Deaf Smith EC provided power to the Hereford Military Reservation and Reception Center, occupying eight hundred acres of land in Deaf Smith and Castro Counties a few miles south of Hereford. The camp housed some five thousand Italian soldiers, most of them captured in North Africa, as well as about 750 US military personnel.[4]

In the fall of 1945, with the end of the war in Europe and in the wake of unspeakable concentration camp atrocities, American attitudes hardened toward German and Italian POWs. Aware of the shocking photos of bodies and skeletal concentration camp survivors, the army colonel in charge at Hereford took it upon himself to drastically reduce both privileges and rations. The POWs began to lose weight and get sick. Some were reduced to eating grasshoppers and rattlesnakes, maybe even stray dogs and cats.

Conditions eased a bit after the bishop of Amarillo, Father Laurence J. FitzSimon, wrote a letter to his congressman detailing the harsh conditions. He reminded the congressman that the Italian POWs had not committed war crimes, were not criminal Fascists, and deserved to be treated humanely.

For a select few, conditions eased dramatically when the Reverend John H. Krukkert, the newly appointed pastor at St. Mary's Catholic Church in nearby Umbarger, discovered that gifted artisans were among the prison population. He hit upon the idea of enlisting the POWs to decorate the parish's drab and unimaginative sanctuary. They refused initially, until Krukkert told them, "I can't pay you, but I can feed you."

Five days a week for the next six weeks, eleven POWs—nine artists and artisans and two helpers—climbed into the back of a truck and rode the twenty miles to Umbarger. They repainted the dingy white walls a cheery pale yellow, with complementary mauve trim. Reflecting their Italian Renaissance roots, they carved a magnificently detailed bas-relief of Da Vinci's *The Last Supper*. They installed stained-glass windows and painted intricate, historically accurate ornamentation on the walls as well as large murals of the Annunciation and the Visitation; both murals featured pastoral Panhandle settings. Their pièce de résistance was the Assumption, a painting eight feet high and twelve feet across of Mary and accompanying infant angels.

The POWs also increased their calorie intake. After working all morning, the men sat down at long tables in the church basement to home-cooked roast sausage, ham, fried chicken, fresh-baked bread, sauerkraut, and mashed potatoes. In the afternoon, they took a break from work for cakes, pies, cobblers, and cookies. They began to regain the weight they had lost on their daily ration of one salted herring or a bowl of watery soup.

Umbarger, Praha, and Frydek were different only in detail during the war years. According to the War Department, no other state contributed a larger portion of its population to the war effort. More than 725,000 men and women served in uniform, including 20,000 Texas Aggies. Nearly three-quarters of the Aggies were officers. Six received Medals of Honor.

Twelve thousand Texas women served, more than in any other state. When Oveta Culp Hobby, a Texan herself, was named head of the Women's Army Auxiliary Corps, she insisted that the army draw up a list of 239 specific tasks to prevent women being shunted aside into meaningless jobs. Their work became so important to the army's operation that the word "auxiliary" was dropped in 1943.

Every city and town in Texas, every family, was affected. Every community sacrificed. Lives were lost (almost twenty-two thousand Texans). Lives were irrevocably changed. Getting married, starting a family, going to school—everyday life itself would have to wait, for the duration.[5]

Less than a decade after the Central Texas town of Bartlett had become the site of the nation's first REA-financed electric co-op, the momentum to bring power to America's farms ground to a halt like a battery drained of juice. Young men who might have been installing utility poles or stringing power lines for the co-op were tramping through malaria-ridden jungles on nameless Pacific islands or flying bombers through flak-infested skies over Germany. Young women who might have been keeping the books and answering phones in electric co-op offices were keeping track of payroll in hastily erected offices at Camp Hood or Fort Wolters or any of the other military facilities scattered by the thousands around the country. Others joined the Women's Airforce Service Pilots (WASP). With male pilots needed for combat, the WASP flyers ferried just-off-the-assembly-line planes from the manufacturers to bases around the country. They trained at Avenger Field near Sweetwater.

With war on the horizon in 1941, an executive order from President Roosevelt in 1942 established the War Production Board. Its purpose was to regulate the production of materials and fuel if war should come, as seemed inevitable. The WPB retooled civilian industries to the production of wartime necessities, allocated scarce materials and services, and prohibited nonessential production.

The REA shifted its focus, as well. With the federal government dedicating most of its labor force and resources to the war effort, the co-ops were barely surviving; the investor-owned utilities (IOUs) contemplated moving in for the kill. E. D. F. Farrow of Hill County EC (now Hilco) joined other co-op leaders from around the country in a decision to get the REA out of Washington. Moving to St. Louis on December 2, 1941, and relying on a skeleton staff, the agency adopted a new slogan—the "Electro-Economy"—and a new mission: providing power to military facilities and related industries. Co-op members would focus on increasing farm production with the help of electricity. They would help meet the nation's defense goals by providing food to friends and allies defending freedom overseas. In 1943, they responded to the challenge and surpassed wartime production goals, producing 14 billion pounds of pork, 11 million head of cattle and calves, and 122 billion pounds of milk.[6]

Immediately after V-J Day, co-ops around the country realized they were running short on the basics: transformers, pole hardware, and other line construction components. A group of Texas co-ops got together and purchased East Texas pole-creosoting plants, but the pervasive bottlenecks continued throughout the war and for some years afterward.

Labor quickly became a problem. The experience of La Grange–based Fayette EC was typical. Founded in 1939 and just getting organized when war broke out, the co-op lost its superintendent, Thomas Hinton, when he joined the navy in 1942. Wiring inspector John F. Luecke joined the navy later that same year. Hinton's wife ran the co-op as assistant manager until April 1943, when Edward A. Giese took over as manager pro tem. Hinton and Luecke returned in 1945. Hinton resigned shortly thereafter, and Luecke took on the task of getting the co-op up and running again. It was not easy.

In 1946, Fayette was swamped with applications but couldn't get materials to build new lines. The co-op could build extensions to serve about fifteen to twenty new members per month.

The electric co-ops had done all they could to assist in the war effort, but, as with Fayette, their basic mission had to wait even beyond war's end. Laborers were not the only shortage; construction materials had also been diverted to the war effort. Electric energy became so scarce in many areas that it was necessary to impose a brownout.[7]

To extend lines into the countryside, for example, co-ops needed aluminum as a conductor of electricity, but the War Production Board declared a freeze on the use of the lightweight metal for anything but bombers and fighter planes coming off assembly lines as fast as the machines could run. Aluminum had always been scarce, primarily because its production requires enormous quantities of electricity. Before full-scale rearmament began in 1940, the aluminum industry had been able to expand gradually to meet the demand for wire by rural electric co-ops, but during the war there was no aluminum to spare. For a while, the co-ops thought copper might be an adequate replacement for aluminum as a conductor, but it wasn't long before copper joined aluminum and steel on the list of raw materials devoted exclusively to the war effort. With production diverted for four years, some twenty thousand miles of cooperative poles were left standing idle around the country, waiting for war's end and the conductor that would light up thousands of farmhouses.[8]

Copper, however, turned out to be a useful weapon in another "war effort"—this one the war the utility industry continued to wage against the cooperatives. Gilmer-based Upshur Rural EC would figure prominently in the skirmish. Upshur Rural, founded in 1937, received an REA loan of $140,000 and promptly built twenty-eight miles of line to serve 139 members. By 1941, the East Texas co-op served 2,100 members in nine counties. That same year, before the onset of war, the co-op borrowed money from the National Youth Administration, another New Deal program, and used NYA labor to build a new headquarters building. It also joined other area co-ops to form Farmers Electric Generating Cooperative, Inc., with the intention of building a $1 million electric generating plant. The new generating cooperative hired contractors to string copper wire from Gilmer to the participating co-ops.

The investor-owned utilities pounced, accusing Upshur Rural of pretending to serve the defense effort in order to hoard copper for its own uses. On December 1, 1941, less than a week before Pearl Harbor, US representative Thomas Winter (R-KS) called for a congressional investigation of the REA, charging that the agency had hindered the defense effort by "improper use of copper while at the same time depriving the farmers of sorely needed copper for the energizing of farm distribution lines." Accusing the REA of "teeming with communists," the congressman showed photos that he said proved the REA "had millions and millions of pounds of copper hidden away in Texas cotton fields."

Here's how the *Dallas Morning News*, the longtime foe of electric co-ops, reported on Winter's charges: "The hiding of twenty-three carloads of copper cable in East Texas cotton fields was cited by Representative Thomas D. Winter (Rep.) of Kansas in a stirring attack on the new deal Monday as proof that 'national defense is being fantastically abused as the excuse for perpetuation and expansion of federal activities which not only are not remotely related to national defense but which divert from bona fide defense activities the energies and resources of the Federal Government.'"[9]

Congressman Bob Poage of Waco, the electric co-op's steadfast champion during decades in Washington, was outraged. Responding at length on the floor of the House, he pointed out that the copper had been ordered for use by proposed generating cooperatives before any restrictions existed. He accused Winter of being a utility industry flack.

"If the farmers of Texas were, as the gentleman from Kansas and the private utility companies state, actually hiding copper they would follow the example of the utility companies and hoard it up in a locked warehouse," the Waco congressman charged. "For many months private utilities over the nation have been receiving deliveries of copper. In fact, most private utilities now have vast stores of copper on hand, and in most cases are telling all prospective customers that they have the material to build any desired lines at a moment's notice. In spite of the fact that very few such private utilities have any available power to sell."[10]

US representative John Elliott Rankin stood with his Texas colleague and other REA defenders. The Mississippi Democrat read into the *Congressional Record* a letter from REA administrator Harry Slattery acknowledging that a million pounds of copper were stored in an open field near Gilmer, but he explained that it had been ordered and received by a contractor months before the Office of Priority Management (later to become the

War Production Board) had restricted its use. "The copper has never been hoarded or hidden," Slattery wrote. The materials, he said, were for transmission lines being delayed because of the shortage of other materials needed for the construction of power lines.

Two congressional committees investigated. Despite Slattery's defense, the House Committee on Military Affairs censured the REA for "planning and projecting many large transmission and generating projects as necessary to the war program, which are not necessary." US representative William J. Fitzgerald (D-CT) filed a minority report that said, in part: "Most of the testimony was irrelevant and was directed not at saving copper but at saving the private utilities from competition."

Five years later, the *Morning News* revisited the story, introducing Upshur Rural's role in the alleged scandal this way: "Between its bumper crop of sweet potatoes and its REA program, Upshur County manages to keep consistently in the national limelight. The sweet potato publicity is always favorable. As for the REA, public reaction is spotty, ranging from fanatical distrust to childlike faith."[11] The story, included in a five-part series on co-ops by Robert M. Hayes, the newspaper's East Texas editor, reflected the *Morning News* publisher's ongoing opposition to electric co-ops. His opposition inevitably spread from the newspaper's editorial pages to its news pages.

In its own pages, *Texas Co-op Power* charged that the *Morning News* "reprinted most of the old arguments against the REA program that the private utilities have been spreading since the program was initiated in 1935. Written in a highly partisan and prejudiced style, the articles quoted frequently from Frank M. Wilkes, president of the Southwestern Gas & Electric Co." *Texas Co-op Power* also assailed the *Morning News* for proclaiming Wilkes "the father of rural electrification in the Southwest."[12]

"Early in 1941," Hayes wrote, "war clouds gathered, and various governmental agencies began a spirited priority race for critical war material. In fact, the race became so spirited that Washington eventually had to issue a freeze order to keep the armament program from bogging down." The alleged hoarding of copper wire in East Texas, with Gilmer at the center of the allegation, created a national scandal, Hayes asserted. He wrote that testimony before a congressional committee revealed that early in October 1941, twenty-three railroad carloads of transmission copper wire—almost a million pounds—showed up in Gilmer and other sites in East Texas and were off-loaded and stored in area cotton fields.

"Witnesses also testified," Hayes wrote, "that in direct violation of a War Production Board order, contractors for the Farmers Electric Generation and Transmission Cooperative, using day and night crews, started building transmission lines the day after Pearl Harbor when there was not even enough copper for the manufacture of small arms ammunition." Hayes further noted in his 1946 series, "The ensuing years have all but erased the memory of the copper scandal," but "plans for the erection of a generating plant at Gilmer have been shelved."

Delta Scales was Upshur Rural's manager from 1944 to 1950, succeeding her late husband, Sam Scales. She had more positive memories of the co-op's activities during the war years. "There were lots of industries in this area that contributed to the war effort, like canning factories," she recalled on Upshur Rural's fiftieth anniversary, "and we had to make sure they got plenty of electric power."

One peripheral, albeit significant result of the "Copper War" was the creation of a national electric cooperative organization. Recognizing that co-ops needed expertise to help them speak authoritatively and in unison—"some sort of mutual protective association," Marquis Childs called it—rural electric leaders from across the nation met in Washington in early 1942 to determine what form an organization would take. "[The Gilmer allegations] and other attacks had created bad feeling," Childs wrote. "The program was in danger of being discredited and thereby brought to a full stop. The same necessities that had prompted the formation of statewide organizations now led inevitably to a national association."[13]

On March 19, 1942, ten of the leaders convening in Washington—farmers, co-op managers, and lawyers—formed the National Rural Electric Cooperative Association. (Hill County EC's Farrow was among the ten.) The group met the next day at the historic Willard Hotel to adopt bylaws, elect officers, and accept memberships.

As the first executive manager of NRECA, the original incorporators chose Clyde T. Ellis, a quick-witted and combative former congressman from Arkansas. A liberal New Dealer from one of the most conservative states in the country, Ellis grew up on a farm near Pea Ridge, Arkansas, without electricity and put himself through college and law school selling Bibles. Elected to the Arkansas House at age twenty-three on a public-power platform, he also served in the state senate, where he introduced a bill that was to become Arkansas' rural electrification act. He was elected to

Congress in 1938 and went to work for NRECA shortly after losing a 1942 race for the US Senate. He soon became a force politicians and the power industry couldn't ignore.

Interviewed by Studs Terkel for his 1970 book *Hard Times: An Oral History of the Great Depression*, Ellis described the dust storms, the drought, the bank failures, and the loss of hope among America's rural people. "Mountain people are more rigorous than others," Ellis told Terkel. "We lived a harder life. We had to grow or make most of the things we needed. . . . We had relatives who just gave up. Broke up homes, scattered to different states. From down in my county, many would go to what we called De-troit. Then they started to go to California, any way they could."

"I'd been talking about electricity for the people," Ellis recalled. "The little towns hardly had any—just a putt-putt plant that wasn't reliable. The rural people had nothing. We could see the fogs rise over the White River, and when the river was up big, we heard the roar. We knew there was tremendous power going to waste. We had read about hydroelectric plants elsewhere in the world. We talked about it, but there wasn't much we could do. The power companies were against it."

Ellis recalled trying to do something about both flood control and power, with the TVA as a model. "We set to form electric co-ops, hoping to buy power from the companies," he said. "It would be too long to wait on the dams. They demanded an outrageous price. They were determined to fight the cooperatives all the way."

Ellis fought back. His rural electrification bill passed, becoming a model for the REA and for electric co-ops in other states. Like Texas, Arkansas began to get co-ops organized in the mid-1930s. "In 1934, I was elected to the state senate, and introduced another rural electrification bill. It was passed in '35, but it was still a struggle. It became a model for the REA and other states used it around the country. In 1936, we got some electric co-ops organized."

By the time Ellis retired in 1967, he was known as "Mr. Rural Electrification." Not long before he retired, he recalled the Upshur Rural brouhaha. "I sat in utter disbelief," he said. "While their charges were broad, it soon became apparent that all they had as evidence was a pile of copper wire with a high fence around it, lying unused, because of the Kellogg OPM order, along the roadside near Gilmer, Texas. The investigations proved the hoarding charges to be completely unfounded."[14]

At war's end, the Czech community of Praha mourned the loss of its nine young men—young men among almost twenty-two thousand Texans who died in North Africa; the Philippines; in the air over Germany, France, and Belgium; in the fathomless depths of the Pacific; and on godforsaken atolls and islands far, far from home. Praha's fellow Texans in nearby Frydek thanked God for the return of their young men. All were eager to get on with lives interrupted. In the Piney Woods of East Texas, Upshur Rural's membership nearly doubled, from 2,672 in 1944 to 4,056 in 1946. It had doubled again by 1952.

In Hereford, the Italian artists and artisans transforming Umbarger's nondescript church into a work of art were reluctant to return home. When the time came for repatriation, they weren't quite finished. Although it had been years since they had seen their families and their native land, they begged to stay three more days. They weren't allowed to stay, but a number of them returned to Umbarger for visits in years to come, as did children and grandchildren in even later years. Several St. Mary's members, proud to this day of their beautiful church, repaid their visits in Italy.

"Some of the boys who are now on the march in Europe and the Pacific will be on the march in this district stringing additional lights in rural electrification extensions after the war is won," Lyndon Johnson told a gathering of co-op members in Austin in the summer of 1944.[15] A year later, the war was over, America was getting back to normal life, and Texas electric co-ops were trying to gin up the energy and focus that had been interrupted by the war. "America just emerged victorious from the costliest, bloodiest, most fateful struggle in history," *Texas Co-op Power* noted in a front-page editorial in its September 1945 issue. "All the material and human resources of our country were mobilized in defense of our civilization against fascism and tyranny. We now face problems of peace scarcely less momentous and difficult than those of war."[16]

The newspaper noted that America had emerged from the war as the richest country in the world, with abundant natural resources, a formidable industrial base, and a trained labor force. It was a combination of advantages that should be able to provide a decent standard of living to every American. "The problem now is to mobilize these same resources, factories and trained personnel for all-out production of the essentials of peace," the newspaper observed.[17]

The war slowdown had created a logjam of applications. Construction began slowly. Hardware, poles, and trained technicians were hard to find, and it was not until 1948 that construction got into full swing. By that year, over forty thousand customers per year were being connected. By June, 78 percent of America's farms were receiving REA service.

Pointing out that less than half of the farms in the richest nation in the world were electrified, *Texas Co-op Power* insisted that every farm in America should have power, and every rural family should have the opportunity to take advantage of labor-saving devices made possible by electricity. "Factories that have been turning out munitions of war should, wherever possible, now be converted to production of a PLENTIFUL supply of the necessities of peace," *TCP* insisted. "Needed on farms are a tremendous quantity of electric refrigerators, stoves, pumps, washing machines, milking machines, radios, motors and hundreds of other items AT PRICES FARMERS CAN AFFORD TO PAY." The newspaper warned that trusts and monopolies stood in the way and were likely to gain control of America's surplus war factories. That must not be allowed to happen, *TCP* editorialized. "All-out production is fully as necessary to win the Peace as it was to win the War."[18]

"WITH ELECTRICITY, THEY HAVE A NEW LIFE"

"If you wanna get religion," George W. Haggard told Washington reporter Liz Carpenter in 1949, "just be present in a Texas farmhouse when they turn on the lights for the first time. It'll make a Christian out of you to see the anticipation in the faces of the men, women and children, who have been living in semi-darkness all their lives, drawing water by hand, washing clothes over a rub-board, reading by kerosene lamp. With electricity, they have a new life."

For Haggard, as Carpenter remarked in her profile for Texas newspapers, the rural electrification cause was, indeed, a religion. "His job as assistant administrator of the REA," she wrote after interviewing him in his office at the Department of Agriculture, "is the perfect outlet for his basic philosophy: To help the people improve their material welfare, and to expand and strengthen the democratic process. The REA cooperatives do both of these things, Haggard believes. They give to the people a new vital resource, electricity, and they let the people have a democratic voice in the operation of this resource." Haggard just might be "the happiest man in the city," concluded Carpenter, who in years to come would become press secretary to a First Lady, Lady Bird Johnson, as well as a noted author.[1]

A native Texan, George Wilford Haggard was born on a Comanche County farm near Gustine in 1909 and grew up in Clyde and Cross Plains. (His father was school superintendent in both towns.) He remembered his twice-daily chore of milking the cows. He also remembered struggling to read by the flickering light of coal-oil lamps. Those experiences likely fueled his zeal to make life on the farm a bit easier.

After graduating in 1930 from Hardin-Simmons College (now University) in Abilene, Haggard became a reporter for the *Abilene Reporter-News* and city editor for the *Sherman Daily Democrat* and later taught high school journalism in Breckenridge and Abilene. He might have continued instructing Abilene teenagers on how to craft a "lede" and conduct an interview if not for J. Walter Hammond, a farmer in nearby Tye who also served as president of the Texas Farm Bureau, a farmers' lobbying organization founded in 1933. As Carpenter noted, Hammond decided that the short, bespectacled young man with dark wavy hair just might make a good editor for the Farm Bureau's publication.

Hammond noticed that Haggard was a good reporter with an ability to handle detail; what he might not have known was that the Abilene journalism teacher also had a combative streak and a missionary zeal that revealed itself in print. Realizing that Haggard's fervor was exactly what the Farm Bureau needed in 1943, Hammond hired the young West Texan as executive secretary of the organization and editor of its newspaper.

At the time, the Farm Bureau was concerned about bills pending in the Texas legislature that farmers worried might cripple the Lower Colorado River Authority and public power in general. Among them was a bill that would have, in essence, brought rural cooperatives under control of the state. Another would have repealed the law that exempted farmers from paying tax on gasoline used in farm machinery. Haggard's lobbying acumen and tough, focused columns and editorials helped scuttle both pieces of legislation.[2]

In April 1944, the Texas Power Reserve Electric Cooperative, the state association of the REA, published the inaugural edition of *Texas Cooperative Electric Power*, forerunner to *Texas Co-op Power*. The publication's editor was one George W. Haggard, who had been lured away from the Texas Farm Bureau a few months earlier to become the TPREC's executive manager. Haggard brought with him the same populist fire and focus that had characterized his work at the Farm Bureau.

The new publication, the third among electric cooperatives around the country, was primarily a political tabloid dedicated to spreading the word about the advantages of the co-op approach. Perhaps more important, it was a weapon in the fight to resist the relentless efforts of investor-owned utilities—the Power Trust, the new publication called them—to drive electric co-ops out of business.

"The Baby Is Born," the voice of the electric co-ops proclaimed on its front page, above an article by Charles M. Curfman, manager of the Greenville-based Farmers Electric Cooperative. "The first issue of *Texas Cooperative Electric Power* marks a distinct milestone in the progress to completely electrify rural Texas," Curfman wrote. "With the advent of this healthy and husky baby among Texas newspapers, there will no longer exist that vacant, uncultivated space within the garden of printed and published fact. . . . No longer will the lay member of the R. E. A. Cooperatives be forced to stand by with a feeling of distress when some writer or individual gives lip service to the poisonous propaganda of the Power Trust, wondering just what is the answer to the unwarranted attack on the Rural Electrification Cooperatives. The answer should be found in the columns of this new publication."[3]

On page 2 of the four-page tabloid, an anonymous editorial (written by Haggard, no doubt) was headlined "A Record of Accomplishment." Noting the cooperative goal "to bring electric power to ALL farms in this country," the editorialist warned of "certain trusts and monopolies that do not like the cooperative way of doing electricity. They will be unceasing in their efforts, not only to handicap the program's expansion, but to deny co-ops access to wholesale power at reasonable rates. The next several years will be crucial. Members of rural electric cooperatives must be vigilant to protect their program from such opposition."[4]

In subsequent issues, Haggard would be more explicit. He wouldn't hesitate to call out the enemy. In the August issue—now eight pages, costing co-op members twenty-five cents annually to subscribe—Haggard warned about "the so-called National Tax Equality Association of Chicago," a lobbying organization masquerading as being composed of small businessmen but backed by large corporations, among them power companies. "The association, which is assailing co-ops of all kinds, recently unlimbered its heaviest artillery against the rural electric cooperatives," he wrote, "flooding the country with a vicious propaganda pamphlet attacking the REA."

Farmers likely would be resistant, Haggard surmised, although town and city dwellers, unfamiliar with the workings of the REA, might be vulnerable to the propaganda. They might conclude that taxpayers were paying for the REA. Haggard called this "an utter and vicious falsehood," although "numerous people are under this mistaken impression, thanks to the widespread power trust propaganda." The editor further noted, "As a matter of fact, the rural electrification program has not cost the taxpayers one thin dime," and in fact, "the government has actually made a profit from its transactions with the co-ops."[5]

The early issues of *Texas Cooperative Electric Power* covered more than just politics and the ongoing battle with the so-called Power Trust. Editor Haggard included stories about all-electric homes co-op members were building, annual co-op meetings, FFA and 4-H club activities, and rural development, as well as plans to organize rural telephone cooperatives and co-op hospitals.

He proudly noted that in 1935, only 2.3 percent of Texas farms had electric service; in 1944, that number had increased to 30 percent. He reminded his fellow Texans that in the number of REA cooperatives, the amount of construction taking place, the total number of miles constructed, the amount of money borrowed from the REA, the number of rural homes served, and the amount repaid, the Lone Star State led the nation.

Still, the primary focus of the publication—renamed *Texas Co-op Power* in February 1945—was the ongoing battle with the powerful lobbying arm of the investor-owned utilities. It was "an all-out fight," Haggard wrote.

He enlisted the likes of Robert Montgomery, a progressive economist at the University of Texas who had thought about running for the congressional seat Lyndon Johnson held. "Dr. Bob," as his students knew him, was a vocal advocate of public ownership of all utilities. "In the 1930s in Texas, calling for government ownership of almost anything other than the military or the mail was enough to get you labeled a wild-eyed radical," one of his students, Bernard Rapoport, recalled years later.[6]

Montgomery contended that the power companies, like the railroads, the oil refineries, and the banking industry, were owned by "foreigners," bankers, and financiers in the Northeast. Texas, he liked to say, was "their largest foreign colony." He estimated that out-of-state corporate interests owned more than 99 percent of the railroads in Texas, 83 percent of the oil refineries, and 95 percent of the state's electric power industry.[7]

The professor maintained that the state's electricity rates were millions of dollars higher than they would have been if Texans had access to public power through a public entity like the Tennessee Valley Authority. Only the federal government could help Texas wrest control from the "foreigners."

Montgomery, like Haggard, admired the populist tradition. His progressive politics earned him the ire of conservatives in the Texas legislature and their appointees to the University of Texas Board of Regents. Accused of teaching Communism, he was called before an investigating committee of the Texas House in 1948 and asked whether he belonged to any radical organizations. Yes, he said. He belonged to the two most radical organizations in existence, "the Methodist Church and the Democratic Party."[8] Montgomery worked on atomic-energy issues during the war and wrote for *Texas Co-op Power* occasionally when he came home to Texas and resumed teaching.

Haggard himself, whose organization was also labeled "communistic" and "socialistic," was almost as outspoken as the professor. In the August 1944 issue, the fiery editor warned that the National Tax Equality Association (NTEA) had opened a southwestern headquarters in Dallas and was raising a slush fund of $58,000 to fight Texas co-ops. "The association has carried on a vigorous assault against all types of farmer cooperatives, flooding the desks of the nation's small and large business men with anti-co-op propaganda," Haggard wrote. The organization's attack, he added, had recently taken the form of a "vicious pamphlet, entitled 'To Hell with the Law,' which was widely distributed."[9]

In October, Haggard reported that leaders from various types of farmer co-ops had met in Fort Worth the previous month to plot strategy. They were particularly incensed with what they considered the NTEA's big lie—that co-ops did not pay taxes. Co-ops pay taxes, Haggard noted regularly, and so do their members. "Naturally they don't pay on excess profits, because they are non-profit enterprises," he pointed out. "Fortunately, after paying their light and power bills, members still have enough left to pay their own taxes, and something left over for their families too."[10]

Also in the October issue, Haggard reprinted an NTEA letter that he said went to businesspeople across Texas. It read in part: "The activities of the co-ops, if continued, will destroy other forms of business, and for the purpose of intelligently and intensely prosecuting every available means of placing an equitable part of the tax burden on these co-ops, the National Tax Equality Association has been set up with headquarters in Chicago and various district offices throughout the country." The letter was signed by nearly two dozen Texas businesspeople, including a Fort Worth feed mill owner, a Canadian (Texas) lumber dealer, the owner of a Dallas cotton oil mill, a Lone Star Gas Company executive, two Texas Power & Light vice presidents, and a Vernon cotton ginner.[11]

Haggard roared back. Under an editorial headlined "Co-ops Are the Essence of Free Enterprise," he pointed out that rural electric co-ops were meeting a need that the nation's gigantic power companies had ignored for more than four decades. "These power companies," he wrote, "with their watered stock and their New York control, said it was impractical, and impossible to electrify rural America." He went on, "Today 70 co-ops in Texas are proving how wrong the power companies were. They are successfully serving 92,000 Texas farm families with the lights and power the private utilities denied them." Haggard warned that the NTEA's propaganda war was "a prelude to a forthcoming showdown fight in Congress and the State legislatures to enact laws that would cripple or kill the cooperatives."[12]

In the December issue, Haggard printed a letter that *Texas Co-op Power* had received from Clyde T. Ellis, executive manager of the National Rural Electric Cooperative Association. Ellis was serving in the navy at the time he wrote the letter. "Dear Editor," he wrote, "have just finished reading your November paper way out here on the broad (censored). . . . Enjoyed it all. Like the way you hit hard at the fascist power racketeers. Your paper is a good morale-builder for us who believe in the people."[13] US representative Wright Patman, the longtime co-op champion, got in a few blows himself, labeling the National Tax Equality Association "one of the sinister and vicious lobbies."[14]

In the November issue to which Ellis alluded, Haggard had described traveling across wide swaths of Texas and stopping to visit with co-op members and managers. His trip began in Fort Worth, where he met with the board of directors of the Brazos Electric Transmission Cooperative, an organization composed of seventeen local distributing co-ops. The manager was R. W. Miller, the driving force behind the very first co-op in Bartlett a decade earlier.

From Fort Worth, Haggard headed westward to Baird for a meeting of District 4 managers and then on to the annual membership meeting of the Fort Belknap Electric Co-op at Olney. From Olney, he drove to Seymour, where the municipal plant "has saved the residents of this little city $20,000 annually in reduced light and power bills."

In Lubbock, where area farmers were gathering cotton and harvesting milo, the editor stopped by the headquarters of the South Plains Electric Cooperative, housed in a two-year-old building that Haggard described as "one of the most beautiful, inside and out, that we have seen." From Lubbock, he traveled to Tahoka, home to the Lyntegar Electric Co-op, where a new construction truck equipped with A-frame and hole digger was ready to extend electricity to more farms on the South Plains.

He visited with the manager of the Bailey County EC in Muleshoe and the president of the board of directors of the Floyd EC in Floydada, where he learned that among the co-op's 625 members were 70 owners of irrigation wells. "These wells," Haggard noted, "tapping a huge underground supply of water, are an important factor in South Plains agriculture."

From Floydada, he drove to Memphis, home to the Hall County EC, and then to the Greenbelt EC in Wellington, where he watched a co-op crew use new equipment to sink a five-foot hole in ten minutes. Then it was on to Childress for a visit with the Gate City Electric Co-op and then to Muenster, where he dropped in on a board meeting of the Cooke County Electric Co-op.

Listening to board members discussing the policies and operations of their co-op prompted a thought: "Suppose these co-op members were being served by a private power company with headquarters in New York City—just how much voice would they have in the operation of the business that served them or in the rates they would pay for lights and power, or anything else in connection with the business?" Haggard's ruminations reminded him of Professor Montgomery's observation about Texas being "the largest and most profitable colony Wall Street has." Haggard had a response: "But the co-ops, if they keep on growing, are going to change that, Dr. Montgomery."[15]

At the end of 1947, Haggard left Texas for Washington, DC, where he became an assistant administrator at the REA. At thirty-nine, he was the first Texan to occupy a high REA position, *Texas Co-op Power* noted. "Haggard's services to Texas farmers will not cease because of the new position he has taken in Washington," the newspaper editorialized. "On the contrary, farmers and ranchers all over the country will now receive the benefit of having another able and conscientious public servant in the Nation's capital."[16]

In the spring of 1951, the *Abilene Reporter-News* reported that Abilene friends were urging Haggard to return to Texas and enter the race for congressman-at-large, if population growth as recorded by the recently completed US Census warranted an additional Texas seat. According to the article, friends noted that he already had a nucleus of a statewide organization in the REA co-ops.[17]

On Saturday, June 30, 1951, a United Airlines plane bound for Denver from Salt Lake City crashed into a mountain west of Fort Collins, Colorado, killing all fifty on board. Among the victims were six REA administrators, including Haggard. The six had been attending an annual conference in Salt Lake City and were on their way to Lubbock for a meeting to discuss telephone cooperatives.[18]

"He was my friend and the friend and respected associate of all who worked with him," said Claude Wickard, REA administrator and former secretary of agriculture. "He came to us Jan. 1, 1948, as assistant administrator after long service to the rural people in his native Texas. He was responsible for inaugurating and for the sound development of the rural telephone program, which was so close to his heart. His passing is a cruel blow to the REA program and to all of us in REA."[19]

Haggard left a wife and two children. He was forty-two at the time of his death.

In the years following Haggard's departure from Texas and his subsequent death, the REA's existential battle with the Power Trust continued, but the focus of *Texas Co-op Power* gradually shifted. Haggard's successors still included news from Austin and Washington and still reported on and editorialized about the threat cooperatives faced, but the populist fire ceased to burn so hotly. The publication began to focus more on general news and features from a co-op perspective. The newspaper introduced a classified-ad section and a regular section of women's news, as well as Texas history tidbits, teen talk, news from other rural organizations, and regular coverage of beauty contests, including Miss Rural Electrification. By 1949, circulation had increased 100 percent.

In March of that year, the name Bill Lewis appeared on the *Texas Co-op Power* masthead for the first time. The 1940 graduate of the University of Texas was listed as advertising manager. In the August 1950 issue, he was listed as associate editor. He was named editor in January 1952, a position he held for forty years.

Two years after taking the helm, Lewis noted that fewer than fourteen thousand members received the first issue of the four-page tabloid newspaper "with the somewhat unwieldy title of *Texas Cooperative Electric Power*." On its ten-year anniversary, July 1954, circulation had swelled to more than 140,000. Lewis noted that *Texas Co-op Power* boasted greater circulation than any association publication in the state.[20]

Karen Nejtek, longtime production manager for Texas Electric Cooperatives (TEC), joined the organization as a clerk typist right out of high school in 1978. She recalled that Lewis, her close friend and longtime colleague, was not inclined to be a crusader in the Haggard mold. Dependable and easygoing, he liked to keep things steady as she goes. His *Texas Co-op Power* reflected the daily lives of rural Texans, most of whom had little time for political battles.[21]

Charles Lohrmann, editor of the magazine on its seventy-fifth anniversary, never met Lewis, but his assessment of the longtime editor's focus echoed Nejtek's: "The co-ops' life-or-death struggle with investor-owned utilities evolved into community building through shared information. The magazine's focus shifted to optimizing the opportunities offered by electricity and answering questions about new appliances for the recently electrified farm."[22]

Communications specialist Jessica Ridge noted that from its earliest issues the publication offered safety tips and practical guidance to help readers derive the greatest value from the life-changing innovations that rural Americans had for so long gone without. In particular, thrift achieved through timely maintenance had been a refrain. "Major repairs, or replacements, can often be avoided by proper care of your appliances and by making minor repairs," the August 1950 issue advised, for example. "By these preventative measures, you can realize the fullest efficiency, value, and longevity from household tools."[23]

In years to come, those "household tools" would include high-tech innovations early co-op members might have had trouble imagining. *Texas Co-op Power* readers learned about electronic meter-reading technology, cooperative-provided satellite TV, computer programs designed to simplify and quantify farm and ranch operations, and software that could digitally monitor entire electric systems. In the special seventy-fifth anniversary edition, former editor Kaye Northcott mentioned a 1986 column titled "Memo from Mary." The column explained how new "cellular telephones" allowed you to "make a telephone call from anywhere, your car, the beach, or from a picnic table."[24]

One of the ongoing themes in *Texas Co-op Power* was the exotic evolving into the everyday. In 1971, the talk was of electric tractors. In 1978, it was a solar satellite that would beam electricity back to earth by microwave. The publication reported in 1980 that Senator James McClure of Idaho foresaw electric cars dominating American highways by 2000. The senator's prediction was a bit optimistic, and yet other "experimental" efforts the magazine explored have gone mainstream.

Near Tulia, in 1979, a wind turbine located on a Swisher Electric Cooperative member's farm was helping irrigate corn and grain sorghum fields. In 1980, Lighthouse Electric Cooperative was involved in a solar "power tower" project to help meet the power needs of Crosbyton. The power tower used the sun's energy to produce steam, which drove a conventional turbine. Also in 1980, Elton McGinnes, manager of Southwest Texas Electric Cooperative in Eldorado, told *TCP* about a geothermal resources committee that was overseeing probes into the earth in search of temperatures high enough to generate electricity.

In 2004, *Texas Co-op Power* proclaimed the West Texas town of McCamey the "wind energy capital of Texas." The magazine reported that hundreds of "monolithic metal giants with three-pronged blades" had brought renewed prosperity. "The wind power source will never be capped. There will always be potential," Walt Hornaday of Cielo Wind Power enthused.[25]

Carol Moczygemba, executive editor of *Texas Co-op Power* from 2007 to 2013, recalled visiting five co-ops across the state in 2011 to explore how the introduction of new technology made a significant difference to co-op members. She found that some were relying on the "smart grid," while others were depending on the state's ample wind supply to make their own energy. Still others were using their personal computers to monitor their home energy consumption. "The story was not an abstract, speculative treatise on technology," Moczygemba wrote. "This was real life with real people. The co-op staff and members I met showed me how technology, rather than being intimidating, was something that could make life easier by saving time and money."[26]

In the latter days of January 1992, Texas Electric Cooperative (TEC) members found in their mailbox a strikingly different *Texas Co-op Power*. Beginning with the February issue, the dependable old tabloid had transformed itself into a glossy, full-color magazine on par with *Texas Highways* and *Texas Parks & Wildlife*. Its circulation exceeded that of *Texas Monthly*. In the coming months and years, the magazine would rely on experienced writers and journalists from around the state for stories that focused on interesting people and places, primarily in rural and small-town Texas, as well as news about the state's electric co-ops and the latest advances in electric power. Every issue included "local pages" tailored to the news of each individual co-op. The magazine also developed a lively online presence.

TEC decided to make the switch partly because some of the larger co-ops had decided the tabloid wasn't meeting their needs, Nejtek recalled. Members seemed to approve. Circulation grew from 400,000 in 1992 to 1.8 million thirty years later. By 2020, *Texas Co-op Power* could boast the largest circulation of any magazine in the state.[27]

In the January 1993 issue, Lewis announced his retirement after forty-four years at TEC, forty as editor. He reminisced about some of the celebrities he had covered—John Wayne on the set of *The Alamo*; Dan Blocker, who played Hoss Cartwright on the popular TV series *Bonanza*; and John Connally, the future Texas governor who served as TEC's first attorney. Lewis recalled a late-night session of the Texas legislature when Connally sent him out for burgers and milk for the lawmakers. He also recalled interviewing the not so famous, including an elderly woman who leaned on the bed of her pickup truck and cleaned her fingernails with a butcher knife as Lewis interviewed her. She had lived on her homeplace her whole life and had farmed it alone since her husband's death. She had electric service now and an electric food freezer. "Her voice broke a bit as she spoke of her gratitude for these things and a water pump," Lewis wrote.[28]

William A. "Bill" Lewis died in 2001. The REA recognized the longtime editor and his successors, and the magazine itself, with its highest honor on several occasions. TEC was honored for producing a statewide publication that presented "lucid forthright contributions to electric cooperative objectives." REA's highest national honor was particularly meaningful to Texans, who recalled one of their own each time it was presented. It was called the George W. Haggard Award.

"EITHER WE ALL HANG TOGETHER, OR WE WILL HANG SEPARATELY. AND ON NYLON ROPES"

Texas Co-op Power may have evolved into a community-building enterprise through shared information, but into the mid-1950s—and to some extent beyond—the REA and individual co-ops around the country were still caught up in a life-or-death struggle with the investor-owned utilities. So-called cream skimming and spite lines were the primary tactics the IOUs used at the local level, while full-court-press lobbying in Washington and in state capitals was more of a coordinated effort.

"The private utilities would be willing to give electric service free of charge to every farmer in America and charge up the loss to their city consumers if by so doing they could eliminate the REA," Texas congressman Bob Poage told a postwar conference on rural electrification held in Austin in November 1944.[1]

The Texas legislature convened in Austin on January 9, 1945, for its first postwar session, and *Texas Cooperative Electric Power* editor Haggard warned co-op members to be prepared. "If history repeats itself," he wrote, "vested interests will sponsor a flock of bills to further their own interests at the expense of the people's. Private utilities, always among the most powerful of all lobbies in Austin, are said to be planning several measures aimed at the cooperatives and public power." There was a possibility, though, that lawmakers would turn their attention to the utilities themselves, perhaps even launching a full-scale investigation into "the power trust's efforts to restrict the development of public power at the expense of the war effort, to retard the growth of REA cooperatives, to evade provisions of the Holding Company Act, to control municipal and state politics, and even to dictate the affairs of Texas educational institutions."[2]

The end of the war seemed to ignite even more ambitious spite-line efforts. *Texas Co-op Power* reported in July 1945 that utility representatives from around the state were approaching co-op members who had been with a cooperative for years but had been unable to obtain service because of wartime restrictions. Farmers living off main-traveled roads were the real losers, *TCP* warned: "Utilities have never sought to give complete area coverage, only running their lines into the thickly settled rural sections. Where this is done, it becomes impossible for the co-ops to go in and serve the remainder of the territory, and farmers in the thinner areas are virtually barred from ever receiving electric service from any source." There was no law to prevent utilities from building spite lines into co-op territory, *TCP* noted, but farmers needed to band together and prevent it, simply by refusing any spite-line offers.[3]

"Every farm family that is approached by a utility representative in the months ahead should ask that representative if his company plans to serve the entire area, or only a selected few farms along the main-traveled road," *TCP* advised. "By taking service from a co-op, even though it may, in some cases, mean waiting a few weeks longer, farm families can be sure that their entire neighborhood will be served, that rates will be cheaper in the long run, that service will be better, and that they will own an interest and have a part in the direction of the electric system which serves them."[4]

In postwar Washington, a Texan skilled in homespun harangue was arguably the loudest and most vituperative enemy of the co-ops. His name was W. Lee "Pass the Biscuits, Pappy" O'Daniel, and rural Texans had been listening to him on the radio ever since electric co-ops had brought his mellifluous voice into their farm homes, thanks to electrification. They considered him one of their own, a friend to country folks. He wasn't.

Wilbert Lee O'Daniel was born in Ohio in 1890, grew up on a Kansas farm, and moved to Texas in 1925 for a job as general manager and then president of Burrus Mills in Fort Worth. In 1931, he began hosting a fifteen-minute weekday radio show at 12:30 p.m. that featured fiddler Bob Wills and his band, the Light Crust Doughboys, named for Light Crust

Flour. O'Daniel cared not a fig for country music, but when the part-time barber's distinctive "Texas Swing" began to catch on and sales of Burrus flour skyrocketed, he was happy to whistle along.

The smooth-voiced salesman with slicked-back hair became "Pass the Biscuits, Pappy." All across Texas, farmers were traipsing in from their fields right after noon and small-town housewives were ironing or folding clothes near the radio so they could listen to Wills ("Aaah-ha!") and his Doughboys and to Pappy's own musical compositions, including "Beautiful Texas" and "The Boy Who Never Gets Too Big to Comb His Mother's Hair." They nodded in agreement as the friendly voice delivered homespun homilies about, say, an aged mother, an Alamo martyr, or an old horse. They heeded his cracker-barrel notions about politics, about which he knew nothing.

Dallas insurance magnate Carr P. Collins Sr. got O'Daniel interested in politics. He told O'Daniel that running for public office would sell a lot of flour. Not long afterward, Pappy shared with his radio audience a letter he claimed was from a blind man urging him to jump into the governor's race. He asked his listeners to send him a penny postcard and let him know whether they agreed. On a subsequent show, he announced the results: 54,499 said go for it; three said the job was beneath him.

In 1938, O'Daniel was one of thirteen candidates eager to succeed Governor Jimmy Allred, an ally of President Franklin D. Roosevelt and a friend of the co-ops, as the state's thirty-fourth governor. In addition to O'Daniel, they included Colonel Ernest O. Thompson, chair of the Texas Railroad Commission and former Amarillo mayor, as well as Jim "Pa" Ferguson, attempting a postimpeachment comeback. Thompson, capable and experienced, was the favorite. He didn't realize who he was up against.

Crisscrossing the state in a white bus, O'Daniel railed against "the professional politicians." With his wife, Merle, and their grown children on board, he and his hillbilly band drew massive crowds in towns large and small. He scribbled a song to showcase his campaign centerpiece, a thirty-dollar-a-month pension for all Texans over sixty-five. The author of "Thirty Bucks for Momma" polled more than 573,000 primary votes, Thompson 231,000. Even with thirteen candidates in the field, O'Daniel avoided a runoff with 51.4 percent of the vote.[5]

"In office, O'Daniel dropped the bait and switched," Texas historian T. R. Fehrenbach noted.[6] The new governor unveiled a tax plan, secretly written by his oil-industry allies and the state's largest corporations, that amounted to a burdensome sales tax on the rural and working-class Texans who idolized him. He tried to appoint to key positions either reactionaries or men with no experience in the areas they were to oversee. His pension plan fizzled, mainly because it would cost four times the entire state budget.[7]

When he announced for reelection in 1940, he declared that his platform was the Ten Commandments, his motto the Golden Rule. This time he campaigned in a $15,000 truck with a papier-mâché capitol dome bathed in floodlights and a platform that hydraulically lifted O'Daniel skyward. Crowds across Texas were mesmerized.[8]

He drew five opponents in the Democratic primary, including Thompson and Jerry "Snuffy" Sadler, a member of the Texas Railroad Commission. "I thought by now the public saw through the Light Crust flour salesman, knowing him as nothing more than a radio personality with a smooth voice and some big businessmen behind him," Sadler wrote in his memoir, *Politics, Fat-Cats and Honey-Money Boys*.[9]

Members of O'Daniel's band, including guitarist Kermit Whalen, a.k.a. "Horace the Lovebird," abandoned the governor for Sadler, but that was about the only good news for O'Daniel opponents. He drew 645,000 votes to Thompson's 256,000. Sadler finished fifth out of six.[10]

In the spring of 1941, US senator Morris Sheppard, an East Texan who authored the Eighteenth Amendment (Prohibition), died after thirty-eight years in office. O'Daniel coveted the seat for himself but was concerned about how it would look to jaunt off to Washington just a few months after being reelected governor. Instead, he tapped eighty-seven-year-old Andrew Jackson Houston, the only surviving child of the hero of San Jacinto. The frail old man entered the Senate as the second-oldest member in history, "if not at death's door, standing on the walkway heading up to it," Austin writer Bill Minutaglio observed. Houston conveniently expired two days before the special election.[11]

O'Daniel jumped in, one of twenty-seven candidates in a field that included an ambitious, relatively unknown congressman named Lyndon Johnson. The thirty-two-year-old Hill Country lawmaker outspent O'Daniel 6 to 1 and by midnight on Election Day led the governor by some three thousand votes—a margin that included eleven thousand more votes from Bexar County than there were registered voters. Four days later, after suspiciously late boxes from East Texas were tallied, Johnson found himself a loser by 1,311 votes.

In Washington, Senator O'Daniel was even more of a reactionary than Governor O'Daniel. He found Communists crouching under every federal rock. In the fall of 1947, President Truman suggested restoring wartime price controls to curb inflation. O'Daniel roared that the president had recommended "radical Communistic opiates as a cure for the dire consequences of 14 years of New Deal delirium tremens." Another postwar initiative, the veterans housing program, was "communistic." A number of O'Daniel's Senate colleagues were Communists, the junior senator from Texas claimed.[12]

When a French ship that was docked at Texas City exploded, killing at least 581 people in the worst industrial disaster in American history, the junior senator from Texas suggested that a Communist conspiracy was behind the disaster. "Due to a rather large number of fires, explosions, and railroad wrecks and other disastrous occurrences in this nation lately, all so nearly resembling disastrous occurrences which proceeded our entry into the last war, due largely to communistic underground activity, I believe it is the duty of the senate to conduct" an immediate investigation, O'Daniel announced.[13]

The Tennessee Valley Authority, bequeathed by an Iowa Republican, was a "Communist experiment of the New Deal," the senator proclaimed. "They've had Communists in there running it, and it looks like they're trying to get them confirmed again," he told fellow members of the Senate Environment and Public Works Committee in 1947. "I respect your rugged Texas viewpoint, senator," responded L. T. Wilhoit, a dairyman and chair of the Chattanooga Municipal Power Board. "But I and my business associates don't consider the TVA to be Communistic. It's done a better job of flood control, soil conservation and forest protection than had ever been done before in that valley. It brought a higher standard of living to my dairy market area, brought more business to the valley and developed more profits." O'Daniel acknowledged those improvements, but he had a question for Wilhoit: How was the TVA any different from what the Communists were doing in Russia? "I don't believe anything is being done in Russia like what is being done by the U.S. Government in the TVA," Wilhoit responded.[14]

Of course, if the TVA was Communistic, then the REA had to be too. O'Daniel trained his Communist-hunting antennae on electric cooperatives in 1947, presumably as he prepared for his reelection campaign. His Democratic primary opponent would likely be the electric co-op champion and implacable O'Daniel foe, Lyndon B. Johnson.

Margaret Mayer, a columnist for the *Austin American-Statesman*, immediately noted the irony implicit in O'Daniel's ridiculous REA charge. She quoted a letter from a reader and O'Daniel supporter who reminded her that "O'Daniel had 300,000 more rural families who own radios today than in 1940." Mayer acknowledged that she didn't know where the letter writer got his information regarding the 300,000 rural radios. "Assuming they are correct," she wrote, "we will go further and assume that they were made possible by rural electrification—that is, the REA, which Sen. O'Daniel had described as communistic. How is the senator going to rationalize an appeal to 300,000 or more who are sitting by radios made possible by a communistic system?"[15]

George Haggard, still editor of *Texas Co-op Power*, found nothing remotely amusing in O'Daniel's witch hunts and incendiary rhetoric. "This false and vicious charge by the junior senator from Texas is a studied insult to the 160,000 patriotic, substantial, tax-paying farm and ranch families of this state who receive electricity through the REA cooperatives," he said in a prepared statement.

Haggard said the senator apparently disliked the fact that REA cooperatives were locally owned and democratically controlled by the rural patrons along their lines. "O'Daniel's attempt to smear the REA program," he said, "is apparently the result of one of three motives, or a combination of all of them:

"One, profound and abysmal ignorance of the way in which the electric cooperatives operate and what electricity has meant to rural people;

"Two, the reactionary northern and eastern Republicans, whose darling O'Daniel is, made it a regular practice to denounce everything that is for the general welfare of the American people as 'Communistic';

"Three, the junior senator from Texas is coming up for reelection next year. This looks like an effort to persuade the private utility interests, which hate the rural electrification program, to make a sizable contribution to his campaign chest."

Haggard noted that demagogues like O'Daniel made it difficult for the public and government officials to recognize real Communists. He also noted that O'Daniel had been an REA foe throughout his Senate term. "Is it not surprising to find him now attempting to justify in this absurd manner his votes against the best interests of the vast majority of his rural constituents?" Haggard asked. "O'Daniel's statement will be of interest to all rural families in this state who through the REA cooperatives are enjoying the benefits of electricity for the first time in their lives."[16]

As it turned out, O'Daniel did not run for reelection in 1948. Johnson ran against the popular former governor Coke Stevenson. In one of the most controversial—and consequential—elections in recent American history, the future president defeated Stevenson by eighty-six late-arriving votes from the infamous "Box 13" in the South Texas brush country. Whatever transpired in the 1948 Democratic primary, LBJ made sure he wouldn't suffer a repeat of his O'Daniel humiliation eight years earlier.

As Pappy O'Daniel headed back home to Texas, he held a press conference to introduce a newly compiled list of "communistic" laws and federal agencies. Heading the list was the Rural Electrification Act of 1936.

In the spring of 1946, co-ops and their allies in Congress—including Texans Sam Rayburn, Wright Patman, and Bob Poage—were trying to pass legislation that would allow the REA to offer loan assistance to generation and transmission cooperatives. The two North Texas congressmen, Rayburn and Patman, were particularly interested because the legislation would allow co-ops to build transmission lines from the Red River Dam near Denison to serve members in Texas, Oklahoma, and Arkansas. Lawmakers working on behalf of the private power companies backed an amendment that would have prohibited so-called G&Ts from receiving REA loans.

In a rare speech on the House floor, Speaker Rayburn rose to complain about power-company tactics. "Washington for the last six months," he said, "has been loaded down and seething with utility lobbyists." One of them went about as low as a lobbyist could go, he charged, by enticing congressmen's wives with nylon hosiery.[17]

The lobbyist in question was Ham Moses, president of Arkansas Power & Light, an investor-owned utility. He carried the nylons in his little black briefcase. Rayburn didn't say so—he didn't have to—but Moses was well aware that nylon hose were so rare during the war years that some women took to painting a thin dark line on the backs of their bare legs, trying to imitate hosiery seams. The nylons were rare because their manufacturer, DuPont, had diverted its production to support the war effort. The moment they came back on the market after a four-year hiatus, stores were swamped. Riots broke out among frenzied shoppers (presumably with runs in their long-used hosiery or dark lines down their legs).

Whether any congressional spouses came away with a pair of nylons will always be a mystery. Nevertheless, Moses's scheme failed. The REA could continue to assist generation and transmission co-ops. "It's almost unbelievable what the power companies will stoop to in their effort to kill us off," said Clyde Ellis. Predicting even tougher battles ahead, Ellis counseled against complacency. "Either we all hang together," he said, "or we will hang separately. And on nylon ropes."[18]

Ellis was right. When the Republicans took control of Congress in 1946, they saw their triumph as a repudiation of the New Deal, including the REA. President Harry Truman requested $175 million in additional REA funding, but the House Appropriations Committee saw fit to provide $75 million. When Democrats forced a vote to increase REA funding, most Republicans voted no.

During the Eightieth Congress in 1947, the power industry launched a campaign to persuade lawmakers to slash appropriations for hydroelectric dams, transmission lines, and other electric power facilities. The REA prevailed, but the battles continued. The National Tax Equality Association and its Texas affiliate, the Federal Tax Equality League, continued to label co-ops as socialists, communists, tax dodgers, and "Joe Stalin's little helpers." The REA noted in 1949 that the NTEA was funded by thirty-one of the nation's largest power companies.

One of the lobbying organization's primary strategies—in addition to its propaganda push—was, in essence, a backhanded compliment to the REA. The group found ready allies among "economy-minded" congressmen who maintained that the rural electrification effort was just about finished; thus, the REA program needed to be drastically cut. Utility executives told Congress that the REA had enough money to finish the job of electrifying America's farms. Marquis Childs reported that the REA at the time had $300 million on hand. "If REA will spend $300 million in the next three years," utility executive Grover Neff told a Senate committee, "and if the utilities will spend nearly as much, the job of extending service to the farms of this country will be approximately at an end." Ellis appeared before the same committee with figures showing that 1.5 million farms would still be without electricity when the REA's loaning capacity came to an end. Congress was persuaded and approved the appropriations the REA requested.[19]

As the 1948 presidential campaign entered the home stretch, the Republican candidate, former New York governor Thomas E. Dewey, was already measuring drapes for the White House. Polls showed him with what seemed to be an insurmountable lead. His Democratic opponent looked for help from the nation's farmers.

The sixty-four-year-old Truman, who had grown up on a Missouri farm with no electricity, running water, or plumbing, kicked off his campaign on September 16, 1948, by making an unannounced visit to an NRECA regional meeting in Washington, DC. "I remember the terrific fight which was made to prevent the passage of the rural electrification law," the president said in impromptu remarks. "This organization and rural electrification have brought things to the farmers that they never dreamed of when I was a kid on the farm."[20]

Campaigning from a special train, Truman headed to the Midwest and then dipped down into the Southwest, speaking before large crowds of farmers and ranchers at crossroad communities and rural county seats. On the train with him were Sam Rayburn and Lyndon Johnson, who was deep into his own campaign for the US Senate. In Bonham, Rayburn's hometown, Truman's speech was broadcast to a national radio audience. "Ask Sam Rayburn how many of the big-money boys helped, when he was sweating blood to get electricity for farmers and the people in the small towns," he urged his listeners. "There have been six record votes in Congress on the REA. In all but one of those record votes, only about 12 to 25 percent of the Republicans voted in favor of the REA. . . . I am deeply concerned about what the Republicans would do to the rural electrification program if they could get control of the whole government."[21] Usually someone in the large crowd would shout out, "Give 'em hell, Harry!" Later, Truman remarked, "I never gave anybody hell. I just told them the truth and they thought it was hell."

Campaigning in Minnesota in mid-October with US Senate candidate Hubert Humphrey, Truman again sang the praises of the REA, reminding a crowd in the small town of Mankato that thanks to the REA and the farmers' cooperatives, six out of every ten Minnesota farms had electricity. He promised that his administration would get the final four, as well. "But in order to do that," he said, "you've got to vote for yourselves. You've got to put somebody in the White House and somebody in Congress that will look after your interests. . . . No one here doubts that cooperatives are a good thing. They have been a tremendous boon to the farmer."[22]

The farmers were listening, and on Election Day they responded. Truman pulled off the biggest upset in presidential history, in large part because of the farm vote. In his book *Campaign*, David Pietrusza noted that among the factors that explained Truman's unlikely victory were "fearful Republican farmers who, in the end, proved more farmers than Republican."[23]

Truman kept his promise. By the time he left office in 1953, 85 percent of America's farms had electricity. His successor, as Don Case notes in *Power Plays*, "was a man of remarkable achievement, but he had spent more time drawing up invasion plans for Normandy than promoting rural electrification."[24]

On November 6, 1952, two days after Dwight D. Eisenhower had been elected president over his Democratic opponent Adlai Stevenson, Lyndon Johnson spoke to the electric cooperatives' annual meeting in San Antonio. The senator tried to assuage member concerns about a Republican in the White House. The American people from every walk of life chose General Eisenhower as their president, he reminded them. "I do not believe they voted against rural electrification and soil conservation," Johnson said. "I do not interpret the election as a mandate to abolish regulation of the commodities and securities markets. I do not think the voters expressed any desire to go back on the progress of the past 20 years."

A few months after the new president had taken the oath of office, US representative Kit Clardy (R-MI) took aim at the REA. A first-term representative from East Lansing and a former chair of the Michigan Public Utilities Commission, Clardy sponsored legislation that would double the REA interest rate from 2 percent to 4 percent. "I am opposed to the whole REA program, to the whole goddamn REA field," he said. He labeled it "pure socialism." Although REA allies in Congress beat back Clardy's bill, it was a portent. The new president was not as crude as Clardy, but he shared his basic premise. Eisenhower had said the right things during the 1952 campaign about "strengthening farmer cooperatives which have done so much," but his administration would never be an ally. "Eisenhower just never did seem to comprehend what the rural electrification program was all about," Clyde Ellis wrote.[25]

The administration consistently maintained that the REA was no longer needed in its current form. In February 1959, NRECA invited Eisenhower to speak to more than seven thousand co-op leaders gathered at the DC Armory in Washington for the NRECA Annual Meeting. It was the largest gathering of officials in the history of the rural electric program. Echoing the bill the Michigan congressman had sponsored in the early years of his administration, Eisenhower told them straight out that the rate of interest at which agencies and individuals borrowed from the federal government was not high enough. He called for doubling the REA interest rate, an action that Ellis and others believed would devastate the program.

The fiery Ellis was not happy, to say the least. He thought the president read the speech "as a teacher would talk to a group of errant schoolboys. He was trying to shame the co-op people into agreeing that it was morally wrong to borrow government money, and wrong to pay the 2 percent interest rate provided by law."[26]

Eisenhower's speech was a prelude to yet another sortie against the REA. The president vetoed key legislation that gave the REA administrator final approval over loan applications. The White House enlisted the support of the American Farm Bureau, and the REA came up four votes short in the effort to override. Although Eisenhower was ebullient about his victory, his proposal to double the REA interest rate went nowhere.[27]

Don Case notes that after Eisenhower left office in 1961, he and wife, Mamie, moved to the family farm near Gettysburg, Pennsylvania, in an area served by Adams Electric Cooperative. The former president attended the co-op's twenty-fifth anniversary in 1965, held at the Holiday Inn in Gettysburg. Ellis was the keynote speaker for the business meeting. The two men shook hands and chatted about cattle.[28]

In 1954, Upshur Rural EC and its headquarters city, Gilmer, found themselves in the news again. The Southwestern Gas and Electric Company of Shreveport, Louisiana, sued Upshur Rural, claiming the co-op had no right to serve some twenty-five of its members who lived in an area that had been annexed by the city of Gilmer. The members all had lived outside the city limits when they first signed on with the co-op.[29]

A federal district judge in Tyler dismissed the suit, but in the fall of 1954, Texas attorney general John Ben Shepperd took it upon himself to sue Upshur Rural. Although the suit involved only twenty-five members of only one co-op, the attorney general sought a ruling that would apply to every co-op in Texas.

"Taken over by the state," *Texas Co-op Power* explained, "the suit has been enlarged to include basic interpretations of the Electric Cooperative Corporation Act, especially the section which says that cooperatives may be organized for the purpose of 'The furnishing of electric energy to persons in rural areas who are not receiving central station service.'"

TEC general manager Elmo Osborne said the effect of the suit would be to drive electric co-ops far away from any populated area. They would be forced to sell off close-in lines "at give-away prices," and they would have to abandon members they had served in good faith. Ultimately, REA loans would be jeopardized.

Osborne and other officials saw the attorney general's lawsuit as an existential threat. "If the attorney general is correct in his interpretation of the law," *Texas Co-op Power* explained, "then the cooperatives will not only have to give up any members who eventually are annexed to any towns, but will naturally not connect any new members who are near any towns for fear that these might be annexed."[30]

In April, district judge Jack Roberts of Austin ruled that fifty other electric co-ops that had volunteered to be parties to the lawsuit, approximately two-thirds of the co-ops in Texas, could stay in to defend their fundamental rights. Co-op counsel William A. Brown, assisted by former Texas governor Dan Moody, called the suit an "industry-wide fight." If the co-ops lost, Brown asserted, it would "destroy the rural electrification movement in Texas."[31]

In June, Judge Roberts ruled against an injunction the attorney general had sought at the request of Southwestern Gas and Electric Company. Shepperd wanted the judge to force Upshur Rural to abandon the twenty-five members in those areas it had served before annexation by Gilmer. Roberts said no, although he also ruled that the co-op could not sign up new members in the annexed areas who were not members at the time of annexation. Co-op attorneys called it a "split decision."[32]

After a delay caused by the judge's emergency appendectomy, both sides appealed. And then waited. On April 17, 1956, the Court of Civil Appeals in Austin issued its ruling. Rejecting Attorney General Shepperd's argument totally, the court ruled that co-ops could not be restricted in their competition with commercial power companies and that exclusive territory franchises were contrary to the Constitution and antitrust statutes. Co-ops would be allowed to continue operating the lines and facilities they had originally constructed in rural areas after annexation, and they could add new members in those areas, as well.

"Thus what appears to be the final chapter in what rural leaders have termed the co-ops' 'battle of survival' has been concluded," *Texas Co-op Power* editorialized. "The Attorney General's attack at the behest of a power company—considered the most vicious in the history of rural electrification in Texas—has been repulsed."[33]

Well, not exactly. The state of Texas appealed to the Texas Supreme Court, whose nine justices heard final oral arguments on January 2, 1957. Newly elected attorney general Will Wilson, on his first day in office, announced that the state would relinquish its time to Southwestern Gas & Electric attorney Dean Morehead.

Morehead told the court that the law as enacted in the Electric Cooperative Corporation Act was clear to the point that electric co-ops had no business in towns of more than 1,500 population and that "the co-ops are not above the law." Upshur Rural attorney E. M. Fulton of Gilmer argued that what was termed the right-of-way act clearly and explicitly gave the co-ops the right to operate inside towns.

On February 6, 1957, the high court shocked the co-ops by reversing the lower-court decision. The court ruled that it was illegal for an electric co-op to serve anyone who was not a member and that only persons living in rural areas and not using electricity could become members. The ruling, in effect, upheld the power-company contention that they had an absolute monopoly to serve urban areas.[34]

"Farmer-owned electric systems have suffered their most staggering blow in the 20 years they have been serving Texans," *Texas Co-op Power* editorialized. The newspaper noted that Upshur Rural's only hope was a change in the law. Texas lawmakers, as it turned out, weren't interested. For the next decade, the Upshur Rural decision was an irksome problem for the co-ops, at a time when cotton and corn fields throughout the state featured a different sort of crop: soon-to-be-annexed subdivisions. Farmers and their families were moving into town.[35]

In 1975, the Texas legislature created a Public Utility Commission, and co-ops agreed to a regulatory scheme that made them fully regulated but fixed their service territories at that point in time. From then on, service territories, with rare exceptions, changed only when two neighboring utilities decided to swap territories to their mutual benefit.

Back in 1944, President Franklin Roosevelt's Republican opponents had found a new target for their anti–New Deal ire. During the presidential campaign, they went after Fala, the president's beloved black Scottish terrier. The story, as the Republicans told it, was that on a visit to the Aleutian Islands Roosevelt inadvertently left Fala behind and then spent millions in taxpayer dollars to send a US Navy destroyer to retrieve the little dog.

In a campaign speech on September 23, 1944, Roosevelt pretended to be irate. "These Republican leaders have not been content with attacks on me, on my wife or on my sons," he said. "No, not content with that, they now include my little dog Fala." Not only was the story completely fabricated, Roosevelt had explained, but Fala was in high Scots dudgeon. "Now, of course I don't resent attacks, and my family don't resent attacks," the president said, keeping a straight face while his audience laughed uproariously, "but Fala does resent attacks."

A dozen years later, electric co-op members may have felt resentful too, when their opponents went after their mascot, a lovable cartoon creation named Willie Wiredhand. It was hard not to laugh at the years-long tempest.

Willie Wiredhand was born on October 30, 1950. Andrew "Drew" McLay, an entomologist and freelance cartoonist who worked for NRECA at the time, was his creator. Willie found his purpose in life a year later when NRECA's membership named him the official mascot of cooperatives nationwide.[36]

"We were toying with ideas for an electric cooperative symbol," recalled William Roberts, editor in the 1950s of *Rural Electrification*, NRECA's monthly magazine. "I had tossed out the idea that the symbol ought to portray rural electric service as the farmer's hired hand, which in those days was almost the whole public relations story we wanted to get across. Drew picked up on the idea at my home one night after a couple of beers." Sprawled out on Roberts's living room floor with a sketchpad and pencil, McLay brought to life "Willie the Wired Hand," later shortened to "Willie Wiredhand."

"Everything about Willie was symbolic of rural electricity," NRECA noted. "He was small and wiry; a hard-working, friendly icon with a big, determined smile. . . . His bottom and legs were an electrical plug, and his body was made of wires. His head was a light socket, and his nose was a push button." According to Willie's official NRECA biography, his last name connoted "the never-tiring, always available hired hand to help the nation's farmers."

Willie, bright eyed and smiling, lived up to his name. His face appeared on light bulbs, in co-op publications, on co-op buildings, and in outdoor signage around the country. The popular little guy represented co-op members in Washington and once stood on onstage with then–US senator John F. Kennedy.

The spunky fellow ran into trouble at the tender age of three. As Richard G. Biever tells the story in the May 2001 issue of *Penn Lines*, the electric co-ops initially wanted to use Reddy Kilowatt as their mascot, the same Reddy Kilowatt the power companies had been using since 1926. Reddy's creator, Ashton B. Collins, had licensed his character to the power companies and objected to Reddy being associated with "socialistic" co-ops. Reddy, he said, was reserved for "investor-owned, tax-paying" electric utilities.

Collins, an Alabama Power Company employee, told his lawyers to be ready to sue NRECA if the organization came up with a character of its own that in any way resembled Reddy. When Willie reported to work for the co-ops, NRECA claimed that any resemblance was purely coincidental.

Collins insisted that the two characters were too close in appearance for his comfort.

In July 1953, Collins and a coalition of 109 private power companies formed Reddy Kilowatt, Inc. The group filed a federal lawsuit in Columbia, South Carolina, against a South Carolina electric cooperative that was using Willie in its advertising. Their brief accused the co-ops of copyright infringement and unfair competitive practices. For relief, they insisted that the co-ops kill Willie.

In his *Penn Lines* article, Biever points out that the complainants weren't all that perturbed about any family resemblance between the two mascots; they professed to be concerned that "Willie's poses" would cause confusion among their customers. Willie's attorneys argued that long before Reddy, other animated characters had seen widespread use in the electric industry as trademarks and promotions. In Biever's words, "Testimony revealed that Reddy's handlers had acted like B-grade movie gangsters over the years, using threats of legal action to 'unplug' other spokescharacters such as Arkansas Power & Light's 'The Willing Watt,' Boston Edison's 'Eddie Edison,' Bradford Electric Company's 'Mr. Watts-His-Name' and 'Elec-Tric' of Cincinnati Gas and Electric."[37]

Even though the case sounded like a real-life version of Wile E. Coyote versus the Road Runner, the co-ops took it seriously. "The evidence now seems pretty conclusive that the Wall Street investor control groups in charge of the profit power empires have stepped up their campaign to destroy rural electric co-ops," *Texas Co-op Power* editorialized in March 1956. "In fact, a large segment of the profit power companies—more than 100 of them—have admitted in their suit against us, in the Reddy Kilowatt versus Willie Wiredhand matter, that they are attempting to take over our service areas and our consumers wherever possible."[38]

In August, *Texas Co-op Power* had good news to report: "A United States Federal Court in South Carolina has awarded a clearcut victory to Willie Wiredhand in his legal battle for survival with Reddy Kilowatt." Federal district court judge Harry Watkins found that the names Reddy Kilowatt and Willie Wiredhand were entirely different, and the two figures themselves did not look alike. Therefore, Reddy Kilowatt had no case. Of course, Reddy's legal guardians appealed, to the consternation of NRECA officials. "This is the most vicious thing that rural electric systems have yet encountered," NRECA general manager Clyde Ellis complained. "We're not fighting one or 10 power companies. We're fighting more than 100 of them!"[39]

As it turned out, Ellis had little reason for concern. On January 7, 1957, a three-judge panel of the Fourth US Circuit Court of Appeals ruled unanimously in Willie's favor. Yes, there were similarities between the two mascots, the judges acknowledged, while adding that Reddy "has appeared in thousands of poses doing almost everything possible and in every conceivable activity. The plaintiff has no right to appropriate as its exclusive property all the situations in which figures may be used to illustrate the manifold uses of electricity."[40]

A beaming Willie Wiredhand got back to work for the co-ops. Years later, when some co-ops in Texas and elsewhere considered retiring their stalwart symbol, surveys found that co-op members preferred keeping him on the job.

"IF THEY HAD KNOWN AT THAT POINT HOW LONG WAS THE ROAD AHEAD, THE SEVEN RANCHERS MIGHT HAVE GIVEN UP"

Walter Prescott Webb, the Lone Star State's preeminent historian in the middle decades of the twentieth century, proposed in his masterwork *The Great Plains*, published in 1931, that an "institutional fault line" bisecting the nation represented an environmental shift between ways of life for the people living on one side or the other. The fault line is the ninety-eighth meridian of longitude, running through the heart of North Dakota, South Dakota, Nebraska, Kansas, and Oklahoma before bisecting Texas just west of Austin. "The distinguishing climatic characteristic of the Great Plains environment . . . is a deficiency in the most essential climatic element—water," Webb pointed out. This deficiency, he noted, determines not only the type of plant and animal life capable of surviving, but also human life and human institutions. To the west of this invisible line, all the way to the Pacific slope, the annual average rainfall is twenty inches and the land is basically treeless—with obvious exceptions, of course. East of this line, all the way to the shores of the Atlantic, the landscape is heavily timbered.

As Webb observed, the line brought a temporary halt to westward settlement, as the pioneers learned how to adapt to the Great Plains, or "Great American Desert," as they called the daunting, desolate land that lay before them. On the eastern side of the ninety-eighth, a wooded, relatively wet environment lent itself to cultivation; on the western side, where the water and woods ran out, farming was a chancy venture (until aquifers and irrigation transformed the Great Plains.) The population on the eastern side was one person per square mile in the latter decades of the nineteenth century; on the western side, the population was one hundred per thousand square miles of prairie.[1]

Texas novelist Elmer Kelton, a San Angelo native, once wrote that his father's family had lived west of the ninety-eighth for more than a century. When his father was about seventy-five, Kelton asked him how many really wet years he had seen. "Four," he replied. "They were 1906, 1919, 1941 and 1957. That was little more than one out of twenty. He had seen twice that many dry spells severe enough to be termed drouths. The rest of the time varied between 'pretty good' and 'pretty dry.'"[2]

Once American pioneers crossed the Mississippi and ventured onto the Great Plains, everything they thought they knew about working and living on the land changed drastically. "The ways of travel, the weapons, the method of tilling the soil, the plows and other agricultural implements, and even the laws themselves were modified," Greg Curtis wrote in an article about Webb in the June 1999 issue of *Texas Monthly*.[3]

In his masterpiece, Webb argues that the Great Plains halted slavery in its tracks. The northern system of agriculture, based on "small farms, free labor and a rising industrialism," could adapt to a semidesert, however difficult; the southern cotton culture, relying on forced labor, could not. Thus, slavery was doomed.

A Texan all his life, Webb also pointed out that the southern Great Plains, particularly Texas and Oklahoma, invented the cowboy tradition. The wide-open spaces were ideal for men on horseback herding large herds of cattle. Those men and women brave enough to inhabit the land and settle it were "thrown upon their own resources." Self-reliance was essential.[4]

Technology, including sources of power, had to pause and adapt, as well. While rural electricity took a while to reach rural Texans on the eastern side of the ninety-eighth, it took even longer to cross that formidable line.

Out in Schleicher County, beyond the ninety-eighth, Mrs. Wily Ratliff had a problem. She was known for her fruitcakes, but on a day in late October 1939, she told Fred Gipson, a roving columnist for the *Corpus Christi Caller-Times*, that she was missing a key ingredient. She called it the "wherewithall [*sic*]." She told Gipson it would take fully half a pint to do the job. True, she was the only one in the Ratliff family who ate fruitcake, and she still had a slice of last year's heavy brown creation left, "but it just seems like it's not Christmas unless I bake a fruit cake."

Six years after the repeal of Prohibition, Schleicher County was still so dry that Sheriff O. E. Conner told her she might have to make a surreptitious call on a local resident who had five bottles left of Prohibition-era home brew that he was keeping as sort of a souvenir. "Mrs. Ratliff," Gipson reported, "is afraid no man with the nerve to smuggle in a little holiday cheer is going to want to part with that much out of his bottle."

Gipson didn't report whether Mrs. Ratliff found the "wherewithall" she needed. Instead he shifted to another Schleicher County resident, Edward Ratliff Sr., who may or may not have been related to Mrs. Wily Ratliff. A resident of Schleicher County for more than thirty years, Mr. Edward Ratliff also had a problem, one more serious than Mrs. Wily Ratliff's.

Ed Ratliff had never been without some good hounds to ride after—"when the call of the chase got to pulling too strong," Gipson wrote. A couple of months earlier, someone had thrown poisoned bait into his dog pen, and he had lost three of his best dogs. "I don't know who did it," Ratliff told Gipson. "I don't want to know. Because, if I did, I'd have to hold myself back every time I saw him, to keep from getting my gun and shooting him down." Gipson sympathized. "Some people may not understand such talk," he wrote, "but I, too, have followed hounds that were better in every way than plenty of men I have known."[5]

Ten years after his conversation with dog-lover Ratliff, Gipson published his first novel. It was called *Hound-Dog Man*. In 1956, the Mason County native published his masterpiece, *Old Yeller*.

The Edwards Plateau beyond the ninety-eighth that Fred Gipson often visited on the trail of Texas tales was still sparsely populated. The great distances separating ranch homes made it difficult to reach one of the major goals of rural electrification—area-wide coverage. Marquis Childs, writing in 1951, focused on one success story west of the ninety-eighth, the Southwest Texas Electric Cooperative.

In the late 1930s, seven of the Ratliffs' Schleicher County neighbors—"seven very determined men," Childs labeled them—began to organize a cooperative. Their names were Ed Willoughby, E. C. Hill, W. R. Bearce, F. B. Gunn, T. W. Johnson, J. Forrest Runge, and P. K. McIntosh. The immortal seven knew that in a county with 244,000 sheep and 3,166 people in 1930, they would have to meet with every single rancher living within the county's 1,311 square miles. They also knew they would have to sign up every one of them if they were to have a chance of getting an REA loan.[6]

The ranchers lived in and around Eldorado, a town "named with more romantic flair than prophecy," reads a typed, time-faded inscription beneath a photograph of namesake Gustav Schleicher in the courthouse hallway. "Ranchers and cowboys who got here first very likely fell under the influence of the Mexican neighbors to the south and figured that 'The City of Gold' would be an appropriate tag to wire to their hitching posts. The imaginative person who fixed the name has been forgotten."

Schleicher the man was born in Darmstadt, Germany, in 1823, and came to Texas in 1847 with the "Colony of Forty." Trained as an engineer, he purchased a large piece of land near Llano. After serving in the Texas legislature, he was chosen in 1854 to survey the territory northwest of Bexar County that became Schleicher County.

Early settlers were would-be farmers and ranchers with the fastest horses and mules, Southwest Texas EC manager Don McCormick told the Eldorado Lions Club in 1962. McCormick was eight years old when the county staged its first land rush in 1901. Four years later, he was on hand for the last one. He and his buddies were trying to get to school but were delayed by the crowd at the courthouse.

Verge Tisdale remembered the land rush, as well. It was the night of August 25, 1901, the elderly Eldorado resident told columnist Gipson nearly forty years later. It was close to midnight, and a crowd of some three hundred men were gathered around the rough-boarded shack that served the town as schoolhouse, church, and courthouse. "In work-scarred hands," Gipson wrote in 1938, "they clutched filled-out applications for land claims, needing only the signature of the clerk to make four sections of Schleicher County land their own at $1 per acre, 40 years to pay for it, with interest at 3 percent. Rolled tightly in the application papers was cash for filing fees."[7]

The county judge, the county clerk, a constable armed with six-shooters, the sheriff, and Tisdale, the sheriff's deputy, were inside the shack. A little after midnight, the county clerk opened up. "Like hot angry cattle when the corral gates are thrown open, the mob rushed the door," Gipson wrote.

A six-shooter hammer clicked. Instantly, a dozen other guns were drawn. All was quiet, each man waiting for what was going to happen.

"The first man to pull a trigger is a dead one!" deputy Tisdale warned. The mob settled down. By daylight, the county clerk had signed his name to between five hundred and six hundred applications. The sheriff and his deputy placed them in a vault and kept them under guard until they could be sent to Austin. Schleicher County's first land rush was over. "Some of the young folks may think they have seen some rough play on football fields," McCormick told his Lions Club listeners more than six decades later, "but it was nothing to compare with this land rush and others."

The legislature in 1905 passed a new law providing that land would be sold to the highest bidder, and the land rushes were over. "The first four sections of land that sold for $5 per acre caused a lot of comment, such as 'they can never hope to pay for this land at such a price,'" McCormick said.[8]

Three decades later, in the spring of 1938, Schleicher County rancher Ed Willoughby had seen firsthand how rural electrification had transformed life in the Lampasas area to the east. He had also heard stories of the profound change around Brady when farmers and ranchers stored their kerosene lamps in the barn and switched on electric lights. Willoughby wrote for an REA booklet that outlined what he and his neighbors needed to do to bring electricity to Schleicher County. What he read seemed easy enough. It wasn't.

Willoughby, Schleicher County farm agent W. G. Godwin, and home demonstration agent Margaret Stewart drove hundreds of miles along lonely unpaved roads, their only companions the occasional roadrunner or jackrabbit, sometimes a skittish deer soaring gracefully over barbed-wire fences. The threesome held more than twenty-five meetings in rural schoolhouses throughout the county. Patiently and enthusiastically, they explained the benefits of rural electrification, sometimes to fewer than five people sitting at children's school desks. They used classroom chalkboards to calculate the benefits for their audiences, no matter how sparse.[9]

Some of their listeners were enthusiastic; others listened politely and shook their heads as they closed the front door. The skeptics considered the undertaking impossible, "far-fetched" (maybe a latter-day illustration of "romantic flair"). Some complained it was too expensive. Others were reluctant to allow easements across their property. Still others, as hardheaded as the sheep and goats most of them raised, just didn't see the need to part with the old ways.

In 1939, Ernest Hill convinced the Eldorado Lions Club that rural electrification would be an enormous benefit to county residents. After P. K. McIntosh talked to officials of the Rosebud-based Belfalls Electric Co-op, he came home to Eldorado with an organizational model for a Schleicher County co-op. He laid it out on a table in the Lions Club meeting place. The original seven appointed a committee composed of Hill, McIntosh, and F. B. Gunn, assigning the three the task of persuading their ranching neighbors to sign up. "They experienced the same difficulties as Mr. Willoughby, Mr. Godwin and Miss Stewart," the Southwest Texas EC history reported, "but these setbacks only made them more determined to achieve their goal."

Several of the original organizers made two trips to Pedernales EC in Johnson City and were impressed with both the organizational scheme and the co-op's day-to-day functioning. "After these trips, they decided that nothing less would suit them for Eldorado and Schleicher County," the co-op history noted.

The group contacted the REA. Not long afterward, B. W. Chesser, a field representative for the REA's Region 10 Applications and Loans Division, showed up in Eldorado. A West Texas native and a former public school teacher, Chesser knew the area well. After extensive discussions with the seven, he suggested that the ranchers hire an assistant engineer to assess the number of locations the county would need and where they would be located.

Given the distances involved and the sparse population, Chesser eventually encouraged the Schleicher County group to consider joining with a Kimble County co-op being formed around Junction, eighty miles to the southeast. Representatives from both groups met throughout the summer of 1939, but they could not come to an agreement. Hill contacted Chesser, who convened a meeting of the two groups at historic Fort McKavett. "Mr. Hill made it quite plain that it was impossible for them to work together," the co-op history reported. "Truthfully, he did not want to. He wanted a cooperative headquartered in Eldorado and would not be satisfied with less."

In October 1940, Chesser returned to Eldorado. Driving through the flat, sparsely settled countryside with ranchers Hill and McIntosh, the REA man mentioned that, given the distances they needed to cover, they would be driving forever mapping out locations. The three men hit on the idea of climbing windmills visible on the horizon in every direction. Hill was able to tell Chesser the distance between windmills, and thus the distance between ranches.[10]

Years later, Eddie Albin, Southwest Texas EC's manager of member services, liked to imagine that scene—Chesser toiling up a windmill ladder, holding on tightly as he peered toward the horizon. "See that clump of trees?" Hill might have asked him. "See that windmill? Way over there."[11]

Windmills or not, Chesser kept coming back. The ranchers secured right-of-way easements for building lines, and if they couldn't track down landowners, they paid for easements from their own pockets. With Chesser's assistance, they filled out a loan application.[12]

"If they had known at that point how long was the road ahead," Childs writes, "the seven ranchers might have given up."[13]

The REA's chief criterion was density, and here was a West Texas county that was anything but dense. The newly formed co-op was applying for a loan to build 192 miles of line to serve only 182 members. And those 182 members represented 100 percent of the rural population of Schleicher County. The REA said no; the loan could never pay out.[14]

The Schleicher County seven, accustomed to dealing with the vicissitudes of weather and market prices, didn't get discouraged. Incredibly, they submitted six different applications, one after another. They pointed out to the bureaucrats in Washington that the average income of ranchers in Schleicher County was twice the state average. They noted that almost every ranch house in the county already had running water, plumbing, electric refrigeration, and other appliances. As part of their multiple applications, they also included guarantees from members to pay high monthly premiums. One rancher guaranteed seventy-five dollars a month, others fifty dollars. Another REA expert came down from Washington to verify those guarantees; he found them to be true.[15]

On a June day in 1941, local ranchers Don McCormick and F. B. Gunn left Eldorado for Austin to obtain a charter for the co-op. Between Johnson City and Austin, the two were involved in a car wreck, although neither was injured. McCormick hitchhiked into Austin, arranged for a tow truck to pick up the car and Gunn, and later that afternoon presented the application to the secretary of state's office. Mission accomplished, the two men rode a bus back to Eldorado.[16]

The fledgling co-op re-signed every potential member in Schleicher County, a process that took time. Chesser came back to town, examined the paperwork, and assured the seven ranchers he was going to submit the application, accompanied by his recommendation for approval. Before the application made its way to Washington again, the Japanese bombed Pearl Harbor.

For the next three years, Southwest Texas Electric Co-op was a co-op in name only. Finally, in January 1945, the co-op let its first construction contract and in April of that year began installing 202 miles of line to serve 149 members. Current flowed over the new lines on January 5, 1946.[17]

Eight decades after its founding, the Eldorado-based co-op provided electric service to nearly ten thousand meters within 7,200 square miles spread across twelve counties, all beyond Walter Prescott Webb's daunting ninety-eighth meridian. Most of its members were still farm and ranch families. Other members included oil-field operations and wind and solar farms. Many of the vast ranches had been subdivided and sold off.[18]

General manager William Whitten, who went to work for Southwest Texas in 1976, chuckled at ranching etiquette. Whitten—known as "Buff," after Buffalo Bill Cody—explained that old ranchers didn't normally make an issue of it, but they considered a two-thousand-acre spread or a five-thousand-acre spread, perhaps purchased by a Dallas or Houston doctor, lawyer, or stockbroker, barely a ranchette. "What some people call a ranch isn't," he said, smiling. Although ranchers still raised sheep and goats across the flat, mesquite-choked region—mohair was the primary product—many of the smaller operations catered to hunters. With a huge population of white-tailed deer and Rio Grande turkeys, "hunting is king," Whitten said.[19]

If, as historian Webb maintained, the unique conditions of the Great Plains prompted a pause while would-be settlers assessed what it would take to survive and thrive in a semiarid environment, then farmers and ranchers on the South Plains around Lubbock were still pausing into the 1930s. Some of them, at least. They were dubious when L. E. Countess of Idalou, C. Z. Fine from Posey, C. E. Lilley from Slaton, M. H. Vardeman from Roosevelt, W. M. Ross from the Southwest Ward community, and other Lubbock County farmers and ranchers came around beginning in 1936. They hoped to sit a spell and extol the benefits of electricity on the farm.[20]

One farmer, who had built a house with no front door so as to discourage visitors, declared that electric lights would cause eye trouble. "Anything going on and off sixty times a second is bound to ruin your eyes," he would declare to anyone willing to listen. He finally had one outlet installed in his house for his radio. Neighbors suspected he might be a bit hypocritical; they thought he also might be plugging a lamp into the outlet.[21]

Some farm families simply didn't have the money to wire their homes, while others doubted they could afford the monthly bills. A few kept using home electric plants and carbide lamps.[22]

Bug Dycus was eleven when the REA brought power to Lorenzo, near the Dycus family farm. "When the REA program came, Daddy liked it," he recalled years later. "And so did every farmer out here. But there were very few dollars. To get the REA off the ground, you had to have five bucks.

Daddy, Clarence Lemons and Mr. Cherry would get in the car to visit people about signing up. They would say, 'Well, we just don't have $5.' But they had four or five hogs out there; Daddy said to sell one. They had an extra old cow; sell her. And that's how the REA got started. From the grassroots. I mean the deep grassroots, because $5 was a lot of money."[23]

A. B. Allen, holder of one of the two original memberships in South Plains EC, told *Texas Co-op Power* in 1957 that whether to join or not to join almost caused a split among families. "The wife wanted electricity, and the husband thought it was a waste of money," he recalled. Allen also recalled that it was difficult to meet the REA's requirement of at least three members per mile.[24]

He remembered that suppliers in the early days thought the rural system was "a plaything" and foisted off on the co-op the cheapest equipment possible. Fuse disconnect boxes had holes, and sparrows would stick their heads in and get electrocuted, throwing the whole line out. Since the birds fell out of the hole, crews had trouble finding the cause.[25]

Despite the skeptics, co-op supporters pressed ahead. The new co-op came into being on March 8, 1937, when the Grange held a countywide meeting in the Lubbock Junior High School auditorium. Seventy people attended. The group appointed a temporary committee to work toward organizing a co-op. Committee members contacted city officials, the Texas Tech administration, attorneys, engineers, the state Grange master, and many others. Vocational ag teachers were especially helpful, coordinating membership drives in their own communities.

Their April 27 application was a pledge from 1,081 Lubbock County farmers to accept electric service at a time when the county was home to 2,600 farms without central-station electricity. The fledgling co-op proposed building 430 miles of line. The REA's reply seemed to confirm the wisdom of the skeptics. No funds would be available until July 1, the Texans were told, and even then, REA administrator John Carmody warned, the agency was doubtful about plans to supply power for irrigation wells.

In September, the committee got word that applications for loans in Texas were six times the available funds. We can offer $100,000, the REA said. The co-op took it, even though the amount was a fraction of the original request. The committee went back to the county map and cut out large sections of the planned lines. On November 27, 1937, the state of Texas issued a charter to the South Plains Electric Cooperative, doing business out of a basement room in the Lubbock County Courthouse and serving

266 members. The co-op had two paid employees. In September and October 1938, the REA approved loans for the second and third sections of lines. By June 9, 1939, South Plains was serving 659 members. By the end of 1940, it had 665 miles of line serving 1,437 members.

S. S. Allcorn of New Deal was a founding director of South Plains Electric Cooperative before becoming president. His granddaughter, the late Myrtle Stringer of Lubbock, went to live with her grandparents when she was eleven and years later recalled the co-op's early days. "My granddaddy got involved with the cooperative, because it was community interest and benefit," she said. "Anything that would help the community, he was for it and would help with it. And when he got started with it, he was just gung ho. He would leave sometimes when it was just barely daylight, and it would be after dark when he came in. He wore out three cars visiting with people, trying to get them to pay their $5 to sign up." She went on, "After our house was wired and hooked up to the electric line, the electrician said, 'Seth, this is your job now. You turn the switch on for the first time.' So granddaddy turned it on and said 'Whoopee!'"

Twenty years later, Allcorn told *Texas Co-op Power* he had pioneered in a number of new communities before moving back to Lubbock. "I had always moved to a community and helped organize the school and church," he said. "When I moved back to civilization, I helped organize the co-op. I feel it has had the most far-reaching effects of anything I have ever done."[26]

In Hudspeth County in far West Texas, within sight of the spectacular Guadalupe Mountains, the remnants of an ancient shallow lake from the Pleistocene epoch glisten in the sun. From US Highway 62, the silvery-gray surface looks densely packed, like the Bonneville Salt Flats in Utah. It's not. Motorists occasionally drive onto the flats, assuming the surface is similar to a hard-packed beach, and almost immediately bog down in the salty muck.[27]

The Pleistocene epoch was approximately 1.8 million years ago, but the salt flats have a more recent history worth remembering. Only a few relatively short centuries ago, in the 1700s, it would not have been uncommon to see a train of sixteen cottonwood carts pulled by sixty yoke of oxen making its creaking, tedious way across the arid wastes of far West Texas. The eighty or so men driving the stolid animals were *salineros*, salt gatherers from the Spanish settlements of Ysleta, Socorro, and San Elizario in the El Paso Valley.

Nearly all the residents downriver from El Paso del Norte, most of them farmers and livestock grazers, were Mexican in language, ethnicity, and culture. Salt was integral to their daily lives. Not only did they rely on it to preserve meat and cure hides, but they also sold it to silver miners around Chihuahua, who used tons in the refining process. And they sold salt to the US Army.

Under Spanish law, the salt beds were common property. After the Mexican-American War, they became unclaimed lands under American law, available to anyone who filed on them. The Mexicans—who became Mexican Americans when the Rio Grande changed channels in the 1820s—believed that everybody had a right to the salt, a right guaranteed by the Spanish Crown centuries earlier and affirmed by the Treaty of Guadalupe Hidalgo. Enduring the heat and the threat of Apache attack in their treks to the salt flats, they never thought to file claims.

They didn't realize that after the Civil War and the chaos of Reconstruction their world was changing drastically. The late Paul Cool, author of the definitive history of the Salt War, put it this way: "If salt was the excuse for war, the underlying reason was this struggle between the rights of the community and those of hustling individualists."[28]

The hustlers were El Paso business owners attempting to acquire title to the salt deposits and then charge for the resources. Among their number was Charles H. Howard, a former Confederate Army officer who had been elected district attorney.

In September 1877, Howard arrested two San Elizario residents heading for the salt beds. An angry mob captured and held the district attorney for three days, and for the next couple of months a war raged between Howard and a handful of Texas Rangers on one side and a mob of several hundred locals on the other. A number of San Elizario residents lost their lives—as did Howard, at the hands of the mob—before an uneasy peace was restored.

San Elizario, with a historic Catholic church at its heart, evolved into an unofficial art colony, one of the three original Spanish towns along the Rio Grande that make up the city of El Paso. A hundred miles due east is Salt Flat, a ghost town near the salt flats, Guadalupe Mountains National Park, and the northern boundary of the Rio Grande Electric Cooperative.

Founded in 1945, Rio Grande EC is geographically the largest electric co-op in the nation. The co-op sprawls across thirty-five thousand square miles and serves fifteen counties and two military bases, Fort Bliss near El Paso and Laughlin Air Force Base outside Del Rio. It also serves two national parks, Big Bend and Guadalupe Mountains. Rio Grande EC even spills into New Mexico, home to three hundred members, and maintains two active interconnections with Mexico. With 1.4 meters per mile, Rio Grande's density is the lowest in the nation.

The co-op is headquartered in Brackettville, a town thirty miles east of the river for which it was named. Brackettville grew up adjacent to Fort Clark, no longer a fort but still in existence as a residential development that takes advantage of the compound's historic limestone buildings and its flowing natural springs, life-sustaining for eons. Brackettville is also home to the legendary Black Seminoles, soldiers who were part African American, part Native American and worked as trackers and Indian fighters in the years following the Civil War. Their descendants still live in and around Brackettville.

Sprawling across much of southwest Texas, from the Mexican border and into New Mexico four hundred miles to the north, Rio Grande EC is a prime example of the delayed push to bring cooperative electricity to ranching country, where the nearest neighbor is likely to be miles away. Given the rugged terrain and vast distances, organizing an electric co-op in a region where coyotes outnumbered potential co-op members wasn't easy.

Chris Lacy was familiar with vast distances. A Waco native, he spent his boyhood summers cowboying on his grandfather's fabled Kokernot 06 Ranch, a massive spread spanning three West Texas counties, larger than the state of Rhode Island. After graduating from Texas Christian University, where he majored in business management and played football for the Horned Frogs, Lacy took over as ranch manager. Serving as a Rio Grande EC director from 1997 to 2006, he was elected board president at various times.

"The first thing I remember about the ranch getting electricity from power lines is just how quiet things got at night," he recalled in a 2017 interview with *RE Magazine*. "Before that, we had generators. It was pretty noisy until the last light was turned off, and the generators would stop."[29]

Lacy's grandfather Herbert Kokernot Jr. was the grandson of David Levi Kokernot, an immigrant from Holland who arrived in Texas in 1831, when it was still part of Mexico. After fighting under Sam Houston in the Texas Revolution, the elder Kokernot registered the iconic 06 livestock brand in southeast Texas in 1873. The 06 stands for his grade rank as a captain in the Texas Navy in the early days of the short-lived republic. The family developed a series of ranches before settling near Alpine in 1882.

"When Rural Electric Administration lines were put in, everybody thought that was a wonderful thing," Lacy told *RE*, recalling when power lines near the ranch were first energized in 1956. "My grandfather worked with Rio Grande Electric Cooperative to get lines extended to the Alpine-Fort Davis area, and we've been members ever since."

The 06 is known for doing things the old-fashioned way, honoring the Old West ranching tradition. Much of the 550,000 acres remains undeveloped. Many jobs have not changed in a hundred years. Still, the Kokernots weren't so tradition bound as to ignore the value of electricity. Refrigeration, air compressors, and commercial-grade appliances are an integral part of life and work on the ranch.

"Where Rio Grande EC has extended lines, it's given us opportunities to hook up wells and pump water for our stock," Lacy said. "Being able to plug equipment in and have power makes it easier to run lights to cattle chutes, scales, and many other jobs."[30]

Dell City, more than four hundred miles northwest of Brackettville, is the nearest town to the aforementioned salt flats. Electricity finally came to the community on the New Mexico border a decade after Rio Grande EC was founded. The town, population a little over three hundred in 2021, was incorporated in 1948, shortly after oil drillers came up dry but accidentally punched into an underground aquifer. Water in this arid part of the world immediately attracted farmers from Texas and New Mexico. A nursery rhyme, "The Farmer in the Dell," inspired their new town's name.

On May 16, 1955, Rio Grande EC's manager, Thomas D. Hurd, threw the switch that brought central station power to Dell City via the co-op's new transmission line from El Paso. It had taken local farmers and ranchers seven years to get power. "Dependable central station electric service for the area now means the end of kerosene lamps and the numbering of days for diesel-powered generators which have done their bit to brighten homes and businesses since the late Forties," *Texas Co-op Power* reported in July 1955. At the time electricity arrived, Dell City still had no telegraph or phone communication, no bus, truck, or rail connections. "But with the coming of Rio Grande Co-op power, the town's citizens are looking forward to rapid growth for their community, which is now emerging from the frontier stage," *TCP* reported.[31]

Actually, Dell City has remained in its frontier stage, losing population every decade since the coming of electricity. Ironically, the fourth-largest user of electricity in the Dell City area—behind three large ranches—is

Blue Origin, the launching site for multibillionaire founder Jeff Bezos's forays into the final frontier.

While Bezos launched crewed rockets into outer space—with Rio Grande's help—the co-op also continued to serve one of the last Old West frontiers in the nation. Its western border takes in Big Bend National Park, a spectacular mélange of mountain and high desert in one of the most remote regions in the continental United States. *Texas Co-op Power* in 1954 called Big Bend, established in 1944, "our secret national park."[32]

When it came to getting electricity to the isolated park, it might as well have been secret. Its two entrances are about eighty miles south of the small Brewster County towns of Alpine and Marathon. Hurd, Rio Grande's first general manager, once gave directions to an Alpine visitor inquiring about how to get to the park: "Go down the street here, turn right, go 70 miles and turn right again. You can't miss it."[33]

The WPA Guide to Texas, originally compiled in 1940, offered a colorful description of a travel route through the Big Bend at about the time Rio Grande EC was coming into existence: "The route traverses the southern part of Brewster County, largest in the State, with an area of almost 6,000 square miles. Yet so thinly populated is the area that rarely are more than 30 votes cast in the lower Big Bend. Neighbors travel a hundred miles or more to attend dances, barbecues, or fish frys. . . .

"Isolation is as great a barrier between the Big Bend and the outside as the jagged mountain ranges. Only the hardiest of men and women brave the loneliness of desert ranches, which range in size from a thousand to a half million acres."[34]

Rio Grande EC penetrated the isolation in 1954, building fifty-four miles of line from a point near Alpine to a new substation near the main gateway to the national park. From there the co-op built twenty-five additional miles of line to park headquarters at the base of the Chisos Mountains and then to the villages of Boquillas and Hot Springs along the Rio Grande.

The site of another challenging project the co-op developed a few years later could best be described as the faux frontier. In the late 1950s, motion-picture star John Wayne was longing to make a movie about the Battle of the Alamo. He was persuaded to rebuild the old mission and Old San Antonio, circa 1836, on a twenty-two-thousand-acre sheep, goat, and cattle ranch north of Brackettville. Wayne produced, directed, and starred as Davy Crockett. Richard Boone, Richard Widmark, Chill Wills, Frankie

Avalon, Laurence Harvey, and Linda Cristal were part of Wayne's star-power constellation. Rio Grande EC supplied the electric power.

Wayne had planned to make the movie in Mexico, until the Daughters of the Republic of Texas got wind of his sacrilege. "Mr. Wayne, you make that movie in Mexico," they warned him, "and we'll shut down the doors of every theatre in the state. It will never be shown in Texas."[35]

Brackettville residents molded more than five hundred thousand adobe bricks for the buildings of Old San Antonio. The co-op furnished power for construction and for the production. An artificial stream flowing through the Alamo relied on an electric pump.[36]

"Electricity is a necessity, but it cannot be allowed to be visible in the movie," *Texas Co-op Power* noted. "So all signs of power lines—and of any other discordant item—must be camouflaged."[37]

Despite a series of mishaps during the shooting—including the West Texas sun melting the glue that held down Wayne's toupee—*The Alamo* was a success, if not the blockbuster the Duke envisioned. The picture earned seven Oscar nominations, winning one, and was 1960's fifth highest-grossing film.[38]

Alamo Village served as a set for more than a hundred different productions after *The Alamo* and for many years was a popular tourist attraction. Run-down and infested with hives of bees and no tourists, it bit the dust forever in 2018. Rio Grande EC continued to serve its vast territory.

"THE DAMNED THING WORKS!"

In the late 1940s and early 1950s, rural Texans ventured down to the hardware store and splurged on a television set. The twelve-inch screen built into a cabinet, complete with either an indoor antenna called rabbit ears that sat atop the TV, or, a few years later, an antenna on the roof of the house, might draw in three or four channels if the owner lived near a city; at other times it might draw nothing but snow and static. The new TV owners probably wouldn't have known it, but the miraculous new appliance occupying a prominent place in the parlor was the brainchild of a Mormon farm boy with the unlikely name of Philo Taylor Farnsworth.[1]

In the summer of 1921, young Farnsworth was driving a horse-drawn disk harrower across a potato field on the Farnsworth family farm near Rigby, Idaho. Texas farmers could no doubt identify. Many of them had stared out from a young age at arrow-straight rows stretching toward the horizon, usually while dragging a cotton sack or wielding a hoe in a springtime field of corn.

The fourteen-year-old, born in 1906 in a log cabin without a phone or electricity, was no doubt daydreaming, mesmerized by the patterns in the field stretching out before him. As he worked—and dreamed—the youngster had an idea, an idea that would seem almost inconceivable to anyone but the most accomplished electrical engineer, an Edison or a Bell. Why, he asked himself, couldn't pictures be transmitted line by line, one line following another like those furrows he was plowing, and then displayed so quickly that a viewer would see a whole picture?

"My father gets to looking at all these lines in the field," his son Kent Farnsworth explained years later, "and he pictured an image with these lines. So, you take an electron down across an image, constantly recording how bright or how dark it is. Again and again until you've done the whole field. And, bam! You come back and you do it all again. Real, real fast."[2]

Young Philo sketched out his notion on a blackboard for his high school science teacher. He would trap light in an empty jar, he explained, and transmit the light one line at a time on a magnetically deflected field of electrons. The result would be an electronic video. Farnsworth's teacher was surely aware that the country genius standing at the blackboard had already taught himself about electricity—and physics—simply by reading every science magazine he could get his hands on. He built his own laboratory in the Farnsworth attic, and at the age of twelve he had repaired a Delco generator on the farm while all the adults were standing around wondering what had gone wrong. He built motors out of spare parts and used them to run his mother's washing machine and other farm machinery. He continued working on his video idea after enrolling at Brigham Young University in Provo, Utah.

In 1926, Farnsworth was nineteen and newly married. His eighteen-year-old wife, Elma "Pem" Gardner, would become his collaborator. She did technical drawing, learned spot welding, and typed his lab notes, chronicling the progress of each day's experiments.

In 1927, the couple traveled to San Francisco, where Farnsworth asked a Crocker National Bank official to loan him $25,000 to set up a laboratory. "The banker laughed merrily when a boy pleaded with him to back his crazy idea with dollars," the Associated Press reported, "just enough dollars to enable him to persist in his notion that people could sit in their homes, twirl a radio dial and SEE, while they heard the president speak."

The banker, James Fagan, asked Farnsworth about his experience. "I used to fix the lighting system on our old ranch up in Idaho," the earnest young man told him. Fagan, surprisingly, loaned him the money, and soon the lab became "a cave of miracles," to quote the *New York Times*. The young inventor filed for his first patent on January 7, 1927. On that date, Philo T. Farnsworth, farm boy from Idaho, invented television.

That's the official date TV originated, although the patents could not be granted until the device had been proven to work, or "reduced to practice." After a few months in the warehouse, Farnsworth and his young brother-in-law, Cliff Gardner, a glassblower, had perfected the world's first electronic television camera tube. Farnsworth called it an "image dissector" because it would transmit an image by dissecting it into individual elements and converting the elements, one line at a time—remember the potato field?—into a pulsating electric current. By late summer, the two young men had rigged up an apparatus that they hoped would send an image from the camera to the receiver.

On September 7, 1927, Farnsworth was ready for his first "transmission." He painted a thick, straight line on a glass slide, and Gardner dropped the slide between the image dissector and a hot, bright carbon arc lamp. As biographer Paul Schatzkin recounts in "The Farnsworth Chronicles," in another room Farnsworth and a friend saw the straight-line image "shimmering boldly in an eerie electronic hue on the bottom of one of Farnsworth's magic tubes. When Cliff [Gardner] rotated the slide, everybody could see the image on the receiver rotate as well, clearly proving that they were witnessing visual intelligence being transmitted from one place to another."[3]

Later that evening, Farnsworth wrote in his laboratory journal: "The received line picture was evident at this time." One of his investors, in a telegram to a friend in Los Angeles, described the historic event this way: "The damned thing works!"

For the next twenty years, Farnsworth struggled to market his fantastic invention and to protect his patents from the Radio Corporation of America (RCA), General Electric, and other large companies eager to cannibalize his discovery. In 1934, RCA began demonstrating its own electronic television system.

On an August morning that same year, the Franklin Institute of Philadelphia invited Farnsworth to conduct the world's first full-scale public demonstration of television. People were lined up for blocks when the doors of the institute were opened. Farnsworth set up a camera on the roof of the building and transmitted pictures to an auditorium downstairs. His audience, enthralled, refused to leave. Finally, Farnsworth told his cameraman, "Point the camera at the moon." The next day, newspapers across the country reported that "after smiling down for years into the world's telescope lenses, the mythical tenant of the lunar planet has been photographed for television."

Farnsworth created the Farnsworth Television and Radio Corporation in 1939. Based in Fort Wayne, Indiana, the company was relatively successful until it was sold to International Telephone and Telegraph (ITT) in 1949.

Although RCA had been forced to pay patent royalties to Farnsworth in 1939, the inventor himself never really received his due. He died in March 1971.

Some years earlier, in 1957, Farnsworth, identified as "Dr. X," appeared on the popular quiz show *I've Got a Secret*. While celebrity panelists Bill Cullen, Jayne Meadows, Harry Morgan, and Faye Emerson waited to identify the mystery guest, the studio audience and more than forty million viewers at home were shown two cards. "I invented electronic television," the first card said. And then the second: "In 1922, when I was 14 years old."[4]

Radio, coming into widespread use a decade or so before the REA was founded, was as enticing to electric co-op members as electric lights and washing machines, electric cookstoves and refrigerators. *Texas Co-op Power* reported in December 1949 that fewer than a quarter of farm families across the United States had bathrooms, while 75 percent had radios.[5]

The state's first radio station, WRR (Where Radio Radiates) in Dallas, originated as a fire-station dispatch used to call police and fire precincts. The man who created the station, Henry "Dad" Garrett, was Dallas's electrical jack-of-all-trades. Garrett was the first car dealer in Texas (for the National Automobile and Electric Company in 1901), the man who installed the first electric lights at the State Fair of Texas, and the man who built the city's first traffic signals. In 1921, WRR became the first licensed radio station in Texas and one of the first in the United States.

By the mid-1930s, rural Texans were gathering around the radio, usually a large wooden console in their parlors, and listening to FDR's fireside chats. After the chores, after the milking, after the supper dishes were washed and dried, they could settle in and tune their radio dial, depending on where they lived, to WFAA (Working for All Alike) in Dallas, KPRC (Kotton Port Rail Center) in Houston, WBAP (We Bring a Program) in Fort Worth, or WOAI (World of Agriculture Information) in San Antonio, among the first of a constantly growing stable of Texas stations. They might listen to the screwball antics of George Burns and Gracie Allen; Fibber McGee and Molly; Jack Benny's variety show; *The Adventures of Sam Spade*; *The Aldrich Family*; *Pepper Young's Family*; *The Whistler*; *The Baby Snooks Show*; *The Cisco Kid*; *Doctor Kildare*; *The Fred Allen Show*; *Hopalong Cassidy*; *Jack Armstrong, the All-American Boy*; *The Life of Riley*; *The Lone Ranger*; *The Mercury Theatre on the Air*; *Our Miss Brooks*; *The Shadow*; *Stella Dallas*, and countless others.

Texans who woke up to a rooster's crow likely tuned in to *The Early Birds*, a morning variety show on WFAA hosted by John Allen. At lunchtime (usually called dinner in rural Texas), they would try to catch W. Lee "Pass the Biscuits, Pappy" O'Daniel's popular show on KFJZ in Fort Worth featuring, for a time, Bob Wills and the Light Crust Doughboys. Some may have responded to O'Daniel's call for a listeners' referendum on whether he should run for governor in 1940.

Rural Texans also may have confessed to listening to Dr. John R. Brinkley, originator of the "goat gland transplant" as a sexual rejuvenation treatment. He broadcast from XER, an outlaw station with studios across the Rio Grande in Villa Acuña, evading the Federal Communications Commission with a signal powerful enough to reach Canada. Brinkley claimed his "compound operation" was guaranteed to relieve prostate trouble, sexual debility, sterility, dementia praecox, hardening of the arteries, high blood pressure, diabetes, neurasthenia, epilepsy, and assorted stomach, kidney, bowel, and nervous afflictions.[6]

Arguably much more useful than Dr. Brinkley's claims were the reports from farm and ranch programs that radio stations large and small began airing across the state. Rural Texans came to depend on early-morning weather forecasts, market reports, and ag news from such familiar voices as Murray Cox on WFAA, Cotton John Smith on KGNC in Amarillo, Johnny Watkins on KWTX in Waco, Bill McReynolds on WOAI, Charlie Rankin in the Rio Grande Valley, and George Roesner on KTRH in Houston. By the 1950s, small towns across the state had their own AM radio stations.[7]

Philo T. Farnsworth predicted that TVs would be in widespread use by the late 1930s, and that individual sets would run about $200. He was a bit optimistic on both counts; AM radio was a mainstay into the 1960s. *Texas Co-op Power* reported in April 1947 that the electrical industry was expected to turn out 750,000 TV sets in 1948, compared to 170,000 in 1947 and a mere 6,500 in 1946.

For *TCP*, it was big news in 1950 when an Erath County couple, according to the page 1 headline, "adds television set." The Oscar Fincher family of Route 5, Stephenville, was the owner of one of the first TV sets the co-op served. An accompanying photo showed a woman sitting in a rocking chair, presumably Mrs. Fincher (the story didn't include her first name), about four feet away from the tiny screen. (Stephenville is now served by United Cooperative Services.) The Finchers, with four of a family of five still living at home, said they enjoyed watching TV. They got most of their programs from station WBAP-TV in Fort Worth, although occasionally they pulled in programs from Detroit and elsewhere. Mrs. Fincher told *TCP* that their TV had very little effect on the family electric bill. Occasionally, she said, friends and neighbors dropped by to watch special programs.[8]

Only three stations served Texas at the time—WBAP-TV in Fort Worth, KTXL-TV in San Antonio, and KNUZ-TV in Houston. More were on their way, but the FCC had frozen all new licenses until the agency could sort out frequency complications. Only 109 stations were in operation across the country in 1950, with plans for a TV network on hold until the FCC could determine how to prevent frequencies from bleeding over into each other. Color TV was a possibility, as well.[9]

According to *Texas Co-op Power*, members were enthusiastic. "It's better to look and listen, rather than just to listen," retired Denton County farmer O. A. Peterson told the newspaper in the summer of 1950.

Arthur H. Koen, a dairy farmer in the Colorado City area, recalled that he had first seen a TV in the city a couple of years earlier. "That's what I need down on the farm," he thought to himself. When people asked him why he wanted a TV set way out in the country, he told them: "We're milking cows and can't go any place. It seems the best way to keep up with what's going on in the world."

John Havelka, who lived near Ammannsville in Fayette County, was enthusiastic about the entertainment value. He said he watched TV almost every night and particularly enjoyed ball games and "some of the darndest wrestling matches." Neighbors bearing cold drinks and something to eat dropped by regularly and, in the words of the *TCP* reporter, "his TV set is the center of a swell party." Havelka said he thought his TV was a little expensive when he bought it but soon decided it was worth the price. He said he wouldn't "part with it for anything."

Frank J. Cernosek, another Fayette County resident, said TV offered entertainment for the whole family—ball games and other sports he enjoyed, cooking demonstrations and fashion shows his wife favored. Each of the Cernoseks' five children had their favorites, as well.

Country store owners brought TVs into their places of business. They discovered that curious customers, like moths to a light bulb, just happened to drop by for a look at the newfangled invention. Maybe they stood around long enough to watch the hilarious conclusion of *I Love Lucy*, or maybe they couldn't tear themselves away from the shoot-'em-up western adventures of

Roy Rogers, Gene Autry, Jimmy Wakely, the Cisco Kid, Hopalong Cassidy, or Johnny Mack Brown—and then felt obligated to buy something.

"These rural TV owners report good reception," *TCP* reported. "Some nights there is interference—just as with radio, they point out. It's a disappointment but not in the least discouraging. And so far as operating costs are concerned, owners say their TV sets have caused no noticeable increase in their electric bills."[10]

In 1950, the few rural Texans who owned TVs relied for reception on "rabbit ears," two thin metal tubes that sat atop the TV in an adjustable *V* position. A scant three years later, large aluminum antennas were sprouting from atop farm and ranch houses across the state. TV stations were broadcasting from San Angelo, Houston, San Antonio, Dallas, Harlingen, Tyler, Amarillo, Galveston, Wichita Falls, Texarkana, Fort Worth, and Austin.[11]

"Texas stations really arrived when cables connecting them with eastern centers of television activity were installed," *Texas Co-op Power* reported in 1953. "In addition to local programs produced by their own staffs, they could offer viewers the cream of the audio-video world: drama, news reports by top-flight commentators, educational features, musical shows. The only limitations were those imposed by the budgets of national sponsors and networks."

Four networks spanned the nation: CBS, NBC, ABC, and Dumont. The two largest, CBS and NBC, offered what *TCP* called "top-notch productions like 'American Forum' or the 'Camel News Caravan.' Kids enjoyed 'Kukla, Fran and Ollie,' while drama lovers tuned in to 'Robert Montgomery' and 'Kraft TV Theater,' among others."

"Humor's banner," *TCP* noted, "is carried by stars like Milton Berle and shows like 'I Married Joan' and 'My Little Margie.' For suspense, there's 'Dragnet,' on which no comment is necessary. And for just about everybody there are the nation's top football games every Saturday."

Texas stations were producing their own shows. In San Antonio, KTXL-TV's offerings ranged from variety shows to a mariachi band. KNUZ-TV in Houston had recently begun using "a magnificent mobile unit which permits on-the-spot telecasting." KPRC-TV, also in Houston, paid attention to agriculture. *RFD-TV*, a daily show produced by the station's farm and ranch director, was particularly popular. Harlingen's KGBS-TV made sure Valley fruit growers were apprised of the latest weather reports, while KETX-TV in Tyler was planning programs on soil conservation and interviews with prominent people in agriculture. KRLD-TV in Dallas,

KGUL-TV in Galveston, KFDX-TV in Wichita Falls, and WBAP-TV in Fort Worth also ran programs catering to farm and ranch viewers. KTBC-TV in Austin offered *TV Dude Ranch*, with Cactus Pryor and a hillbilly band.[12]

Decades later, *TCP* associate editor Carol Moczygemba, a self-described "charter member of the baby-boom generation," recalled those halcyon years of early TV. Growing up in Austin, she recalled the excitement that rippled through the neighborhood when the Hendersons next door brought home the first TV on the block. Its arrival changed daily routines for young Carol and her friends. Instead of riding their bikes around the neighborhood looking for playmates, she and her younger brother started hanging out on the Hendersons' front porch. "We would lean into the screened door, our hands cupped around our faces as we waited for our little friend to invite us in to watch Roy Rogers ride Trigger into high adventure," she recalled.

On Sunday evenings, weather permitting, neighbors strolled down the block to the Hendersons', set up their lawn chairs in the front yard, and, with children and popcorn in tow, waited for Mr. Henderson to set up the TV in the living-room picture window, fiddle with the rabbit ears, and turn the sound up full volume. "Then the miracle happened, live from New York, the *Milton Berle Show*."[13]

About the time the Moczygembas and their neighbors were sitting on the Hendersons' front lawn laughing at the antics of Uncle Milty, *TCP* was assessing the potential of the miracle in their midst. "There is no foreseeable boundary to the progress television can make, either technically or in the programs it can offer," the newspaper enthused in 1953. "There is, for any family who buys one of the magic little boxes, new horizons in entertainment and education."[14]

Moczygemba, writing in 1999, was a bit more measured in her assessment of Philo T.'s miracle. By then, 98 percent of American households had at least one TV. Forty percent had three or more. "No longer a novelty," Moczygemba noted, "television has taken on an identity far surpassing its technology. A companion when we're alone, a babysitter when we're busy, a sedative when we're stressed. We turn on the tube when we want to know what's going on."

TV, in other words, had become ubiquitous as the twentieth century became the twenty-first. It was a national habit that, in the days before smartphones, seemed to mesmerize the nation like no other electronic device. "As with any other habit, it takes some effort to step back and question its place in our lives," Moczygemba wrote. "But that's exactly what we should be doing. To question the sorcerer is to break the spell."

Assessing TV's influence on rural America, Moczygemba quoted former TEC general manager Jim Morriss, the first program director for KTBC-TV, Austin's first TV station. Morriss, who died in 2009, was also a former board member of the National Rural Telecommunications Cooperative (NRTC) and the son of Mabel Morriss, the aforementioned Bowie-Cass EC "ramrod." NRTC's mission was to make sure that TV broadcasters didn't ignore rural America.

"The whole idea of NRTC was to provide the same kind of education, stimulation and exposure in areas beyond where cable reached or was expected to reach," Morriss told Moczygemba. "It was based on the same premise as the argument for rural electricity. There's no reason for people to be penalized simply because they live in the country, especially now in this so-called information age."

Morriss maintained that the effect of TV on rural America had been profound. In his view, "it had long since obliterated the line of demarcation between urban and rural."[15]

"I'll get it for you," Congressman Lyndon Johnson promised his Hill Country constituents. By the time he ran for the US Senate in 1941, he had delivered on his promise. Farmhouses throughout his district were wired for electricity. Courtesy Lyndon Baines Johnson Presidential Library; photograph by *Austin American-Statesman*.

Two powerful Texans in Washington, DC, US senator Lyndon Johnson and House speaker Sam Rayburn, shared a rural Texas past. They knew what it was like to live and work without electricity. Courtesy Lyndon Baines Johnson Presidential Library; photographer unknown.

Texas Cooperative Electric Power

News of Interest to Rural Electric Cooperatives in Texas

Published by Texas Power Reserve Electric Cooperative, Inc.

Vol. 1 Austin, Texas, July 1, 1944 No. 1

THE BABY IS BORN

By CHARLES M. CURFMAN, Manager, Farmers Electric Cooperative, Greenville, Texas

The first issue of "Texas Cooperative Electric Power" marks a distinct milestone in the progress to completely electrify rural Texas. With the advent of this healthy and husky baby among Texas newspapers there will no longer exist that vacant, uncultivated space within the garden of printed and published facts.

No longer will the lay members of the R.E.A. Cooperatives be forced to stand by with a feeling of distress when some writer or individual gives lip service to the poisonous propaganda of the Power Trust, wondering just what is the answer to the unwarranted attack on the Rural Electrification Cooperatives. The answer should be found in the columns of this new publication.

There are no facts which the R.E.A. Cooperatives want hidden. "There are no bugs under the cooperative chip." The Co-ops want the full white spotlight of public knowledge turned upon them. They want the public fully informed of what they have done, what they are doing and what they plan to do.

WANT PUBLIC INFORMED

The more the public knows about the R.E.A. Cooperatives, how they function, why they function and what they are accomplishing and the service they are rendering directly to rural people and rural institutions and indirectly to urban people and urban institutions, the more friends and supporters they will have.

There is a great potentially productive assignment awaiting this newcomer to the newspaper family. Great, indeed, is its opportunity for carrying forward a much-needed educational program in the field of Cooperative Rural Electrification. There is much yet that the members of R.E.A. Cooperatives should learn of the many uses to which electric power is adaptable, of the money and time that it will save them, of the conveniences of doing things the electric way. This publication, with its widespread circulation, should furnish the vehicle with which to carry information to the 100,000 cooperative members and their families.

ELECTRICITY MAKES FARM BETTER PLACE TO LIVE

There is a butcher, a baker, a banker and a groceryman in your town who has been wondering why, during the past few years, the farm has so suddenly become such a desirable place to you on which to live. Of course, Rural Electric Power is the answer. You should see to it that a copy of your publication gets into their hands that they may learn of this electric farm economy. With all city conveniences now available, plus the soul-cleansing atmosphere of the country, with elbow room, with a cow and chickens and a garden of their own, they may decide to move to a small farm while continuing their businesses in town.

SOME MISINFORMED REGARDING REA CO-OPS

There are many people in the cities and even some in the country who have been getting their information through the Power Trust propaganda mill, who think R.E.A. Cooperatives a bad thing, with the result that, from time to

(Continued on Page 3)

FARMERS WIN MAJOR VICTORY IN PASSAGE OF PACE BILL

WASHINGTON, D. C., June 24. —Rural electrification won two outstanding victories and suffered one unexpected reverse in recent legislatuve developments at the capital.

One of the greatest victories in years was the approval by both Houses of Congress of Title V of the Pace bill which extends the loan period and reduces interest rates to REA cooperatives.

This measure passed both House and Senate and went to the President shortly before Congress recessed.

Passed with the active support of the national association and the various state organizations of REA cooperatives, including the Texas Power Reserve, it is estimated the Pace bill will save the co-ops a total of $2,500,00 annually in interest charges.

Briefly, Title V of the measure will:

Renew the authority of the REA to borrow from the Reconstruction Finance Corporation. (Present authority expires June 30, 1946.)

Amend Section 4 and 5 of the REA act, providing for interest at the rate of 2 per cent on new and existing loans to co-ops. (Present rates have ranged from 2¼ to 3 per cent.)

Extend from 25 to 35 years the period of loans under Section 4 of the Act.

Amend Section 3(b) to authorize the appropriation of such sums as Congress may from time to time deem necessary for the purposes of the act.

The second victory for cooperatives was the approval by the Senate Commerce Committee of the Bone amendment to the current

(Continued on Page 4)

State Meeting of REA Co-ops To Be Held in Austin July 10-11

Texas Power Reserve Sets Up State Office

The Texas Power Reserve, statewide association of rural electric cooperatives, set up state offices in Austin May 15. Headquarters are in the Scarbrough building. G. W. Haggard, formerly executive secretary of the Texas Farm Bureau Federation, has been named executive secretary and editor of the statewide paper which the association is sponsoring.

Organized in December, 1940, to bring about a reduction in rates of wholesale power to cooperatives, the Power Reserve has a record of outstanding achievement to its credit. At the time of its organization, co-ops were paying 13 to 16 mills per KWH to the private utility companies for wholesale power.

RATES REDUCED

Today, as a result of concerted efforts by this organization and the wholesome influence of the Brazos river project and other public power agencies, rates have been substantially reduced over most of the state, now ranging down to 6 mills in some areas. In general, reductions have averaged around 33 per cent. Further rate reductions can and will be brought about by the united effort of all the cooperatives.

The Power Reserve serves as a coordinating, service, and educa-

(Continued on Page 3)

Chesser Is Named Region 10 Head of Applications, Loans

B. W. Chesser, for the past four years field representative for the Applications and Loans Division of REA in Region 10, has been made regional head of this division, with headquarters in St. Louis, Mo. He assumed his new duties there the latter part of June.

A native of Martin county, West Texas, Mr. Chesser graduated from Texas Technological College at Lubbock in 1931. After teaching school several years he served as AAA administrator in Crosby county and later as county agent of Haskell county for three years. He was with the Soil Conservation for two and one-half years before coming with REA in 1940. For the past two years he has been field chief in Region 10.

Mr. Chesser has many friends among cooperatives in the Southwest, having traveled extensively over this region in the last four years, working with local managers and boards of directors in making electricity available to rural homes.

"My work in the field has been most pleasant, and it is with sincere regret that I leave the state," he said. "However, I feel that this new assignment offers larger opportunities for service to the rural electrification program and to the people of Region 10.

"I expect to be back in the state as often as possible and I will continue to need the same counsel and cooperation from my friends in the cooperatives that I have always enjoyed in the past in order to render maximum service."

Electricity's Place in Food Production And Post-War Will Be Discussed

Cooperative electricity's contributions to wartime food production will be reviewed and its place in post-war agriculture discussed at a statewide meeting of rural electric cooperatives in Austin July 10-11, 1944.

"Food will win the war and write the peace" said Secretary of Agriculture Wickard in the early days of the present struggle, and no factor has played a more important part in increased production on the farm than the labor-saving service of electricity. Likewise nothing is expected to be of greater importance in the post-war period.

WILLIAM J. NEAL

Texas' statewide meeting comes at an opportune time. Still larger food production for our armies and our fighting allies in the crucial days ahead is the first consideration of every farm family in the land—a greater production that can be brought about by larger and more efficient use of electricity.

POST-WAR ASPECTS

Also, before another annual meeting rolls round, America may be in the first stages of its program of post-war reconversion. Rural electrification is expected to play a major role in this, not only because of its vital importance to agriculture, but also as a means of providing employment in the manufacture of electrical supplies and appliances.

These and other aspects of the rural electrification program will be given thorough consideration during the two-day meeting. Managers, employes, directors, and members of every electric cooperative in the state, as well as other leaders in the field of agriculture, are cordially invited to attend. Members of local boards of directors are especially urged to be present and participate in the meeting.

NEAL TO SPEAK

Among the speakers will be William J. Neal, deputy administrator of the Rural Electrification Administration, St. Louis, Mo., who will speak to the group the afternoon of the first day. Congressman W. R. (Bob) Poage will be the speaker at the banquet

(Continued on Page 4)

OVER 200 EXPECTED AT ANNUAL MEETING

As this issue went to press, word had been received from 41 co-ops that they will be represented at the annual statewide meeting in Austin July 10-11 with a total of 119 delegates.

Reports were coming in daily, and attendance at the annual meeting is expected to exceed 200. Co-ops which have not yet reported their estimated attendance to the state office are asked to send these in at the earliest possible date so that arrangements may be completed for the banquet and other sessions.

BRIDGES IN WEST TEXAS

Q. C. Bridges, formerly lineman with the Navarro County Electric Cooperative, has recently taken a position as supervisor of the safety and job training program in the West Texas area.

Your Cooperative Newspaper

This is Volume I, Number 1, of TEXAS COOPERATIVE ELECTRIC POWER, a new statewide publication devoted to promoting the interests of rural electrification in Texas.

The editor and board of directors of the Texas Power Reserve, state association of rural electric cooperatives, which is sponsoring the new publication, believe there is a definite need for a statewide news organ of this sort—to help in coordinating the activities of the various co-ops, to keep members informed of developments, both state and national, that affect them as consumer-owners of rural electric systems, to carry information and suggestions regarding new and more efficient uses of electricity and in various other ways serve as a general medium of information.

Naturally, we expect to enlarge and improve the publication as time goes on. For example, it is planned to add a woman's page, carrying features of special interest to homemakers; also a utilitization page, perhaps a youth's page, and many other features. We especially invite contributions to a "Letters From Readers" column which we hope to start, along with other features, in the forthcoming issue.

This publication is a 100 per cent cooperative enterprise. It belongs to the rural electric cooperatives that make up the Texas Power Reserve. Therefore, it belongs to you, as a consumer-owner of one of these local co-ops.

Keeping pace with the general growth of cooperation in America, the cooperative press has made rapid strides in recent years. True to the spirit of the co-ops, these publications have made service to the readers their first consideration, because the readers own the business, just as they, as consumers, also own the businesses that serve them with electricity.

Rural electric cooperative members of Texas, this is your paper. It will be largely what you want and make it. We hope every cooperative in Texas will subscribe 100 per cent. Criticisms, suggestions, and comment will also be appreciated.

Together, let's work to give Texas the best cooperative newspaper in these United States.

ATTEND STATE-WIDE MEETING OF REA COOPERATIVES IN AUSTIN JULY 10-11, 1944

"The Baby Is Born," the inaugural edition of *Texas Cooperative Electric Power* proclaimed in April 1944. Forerunner to *Texas Co-op Power*, the voice of Texas Electric Cooperatives was feisty and combative in the early days of rural electrification. Courtesy *Texas Co-op Power* magazine.

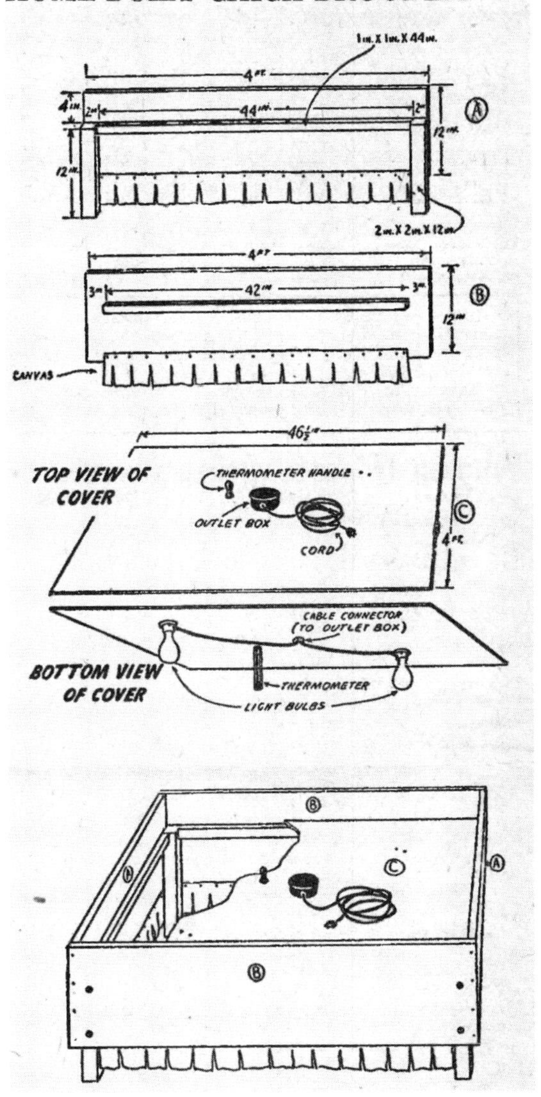

Rural electrification transformed life on Texas farms and ranches. The lights came on, to be sure, but electric power spawned countless time- and labor-saving devices, including a new type of chick brooder for poultry farmers. Courtesy *Texas Co-op Power* magazine.

LBJ, shown with an unidentified woman at the Pedernales Electric Co-op headquarters in Johnson City, stayed in touch with the co-op long after he helped get it organized as a young Hill Country congressman in the late 1930s. Courtesy Lyndon Baines Johnson Presidential Library; photograph by Yoichi Okamoto.

Lyndon Johnson is shown here at a national REA conference accompanied by Miss REA and Miss Texas. Courtesy Lyndon Baines Johnson Presidential Library; photograph by Mike Geissinger.

A *Texas Co-op Power* cartoon in 1950 reminded readers that electricity on the farm made life easier and more efficient in countless ways. Courtesy *Texas Co-op Power* magazine.

Lineworkers have always been the co-ops' representatives in the field, whether they're doing routine maintenance or responding to an emergency. This repair crew, on the job in February 1950 near the small Central Texas town of Mart, worked for Limestone EC. Courtesy *Texas Co-op Power* magazine.

ANNUAL MEETING SET

Hill County Electric Cooperative

THURSDAY, JULY 16th—7:30 P.M.—ITASCA FOOTBALL FIELD

The July annual meetings of the local co-ops have always been a highlight for members and still are today. At its annual meeting in 1964, the Hill County (now Hilco) Electric Co-op gathered quite a crowd on the football field at Itasca High School, "Home of the Wampus Cats." In 2008, members of all ages showed up at the San Bernard Electric Cooperative in Bellville for door prize drawings, barbecue, awards, and election of board members. Courtesy *Texas Co-op Power* magazine.

Annual Meeting Draws More Than 700

The 2008 SBEC annual meeting was held May 17 at the Austin County Fair Expo Building in Bellville. About 715 members, guests and employees attended the event, which began with a business meeting followed by the election of directors and ended with a prize drawing and barbecue meal.

Members, guests and staff all enjoyed fellowship and food at the 2008 annual meeting.

Once electricity finally got to rural households across Texas, co-op members had to learn how to use it. The REA hired home demonstration agents to teach people who might never have used electricity how to operate and maintain electric devices, how to cook and clean with electricity, and how to use the system safely. Pictured is a home demonstration agent assisting a housewife in 1960. Courtesy *Texas Co-op Power* magazine.

Tornadoes, hurricanes, ice storms, and floods, as well as routine maintenance, are par for the course in the workday life of co-op lineworkers, including this crew doing repairs on Lake Texarkana for Bowie-Cass EC in December 1972. Courtesy *Texas Co-op Power* magazine.

Fred McClure, state president of the Texas Future Farmers of America, in 1972. McClure hailed from San Augustine, where his parents were members of Deep East Texas Electric Cooperative, and his profile appeared in the August 1972 issue of *Texas Co-op Power* magazine. He would go on to be elected president of the student body at Texas A&M University and would later serve in the offices of Senator John Tower and President George H. W. Bush. Courtesy Fred McClure.

An integral part of the electric co-op tradition involves reaching out to members. In April 1992, Bartlett EC representatives conducted an education and safety class at Salado Elementary School in Bell County. Courtesy *Texas Co-op Power* magazine.

In October 1984, lineworkers for Coleman County EC were upgrading infrastructure near the small community of Zephyr. Courtesy *Texas Co-op Power* magazine.

The annual Texas Lineman's Rodeo, like a cowpuncher's rodeo, not only highlights skills that come into play on the job but also underscores the hard work and commitment to excellence that are integral to co-op culture. In 2009, Bryan Texas Utilities lineworkers competed at the Texas Lineman's Rodeo in Seguin. Courtesy *Texas Co-op Power* magazine.

TEC's utility pole manufacturing facility, deep in the Piney Woods of East Texas near Jasper, was established in 1946 and has produced more than one hundred thousand premium poles. Professional foresters select only the highest-quality trees, visually inspecting them for straightness, height, knots, scars, and any other defects or diseases. TEC distributes the poles through thirty-three pole yards located around the state. Courtesy *Texas Co-op Power* magazine.

The Electric Reliability Council of Texas manages the flow of electric power to 24 million Texas customers — representing about 90 percent of the state's electric load. In ERCOT's control room, above, the electric grid is monitored constantly. *Photo courtesy of ERCOT*

Bluebonnet keeps an eye on the state's power supply during a blistering summer

Blistering summer heat in 2018 tested the capability of Bastrop-based Bluebonnet EC. One of the oldest co-ops in Texas, Bluebonnet met the test with a commitment to its members and reliance on state-of-the-art equipment. Courtesy *Texas Co-op Power* magazine.

Winter Storm Uri in February 2021 wreaked havoc on the Texas electric grid. Crockett-based Houston County EC met the challenge "through a team effort, a community-based effort," general manager Kathi Calvert recalled. "That's why co-ops are trusted," she added. Courtesy *Texas Co-op Power* magazine.

13

"WE ALWAYS MANAGED TO SHIP OFF THE YOUNG AND ABLE"

In the early decades of the twentieth century, a visitor to Marlin, Texas, could stop by the Missouri Pacific train depot and see people confined to stretchers being brought to the Falls County town, known nationwide for its healing waters. After a month or so of drinking and bathing in the hot, smelly mineral water bubbling up from underground, they were ready to go home again—not on a stretcher but with a skip in their step and a gleam in their eye, eager to spread the word about their miraculous cure.

Marlin boasted hotels (the largest built by Conrad Hilton), bathhouses, boardinghouses, a hospital and disabled children's clinic, and a pavilion where people gathered to soak their feet or take a sip at a flowing fountain of hot, odiferous water. Fifty thousand visitors a year took advantage of the sulfur water's restorative powers.

John McGraw, legendary manager of the New York Giants, swore by Marlin's water. From 1908 to 1918, he held spring training in the little town southeast of Waco. Baseball's "Little Napoleon" believed that the waters were good for sore pitching arms and winter-weakened muscles (not to mention the presumed palliative effects of sulfur on other diseases ballplayers of that era tended to pick up).

By the latter decades of the twentieth century, Marlin itself was in need of healing waters. Once-majestic Victorian houses in pleasant neighborhoods had begun to deteriorate; many were abandoned. As the years passed, smaller houses in and around town fell in on themselves. The nine-story

Falls Hotel—Hilton's finest, where Tommy Dorsey once played for cotillion dances—closed its doors and sat empty over the years. Several light industries left town, and so did the Veterans Hospital, costing Marlin several hundred jobs. Despite efforts over the years to revive the old town—efforts hampered by local bickering, usually along racial lines—the Falls County seat remained a broken shell of its former self. The healing waters remained, but few came to partake.[1]

It was an old story in Texas. Once-vital small towns in every region began emptying out in the years following the Depression and World War II. Paducah was one of them. In the early decades of the twentieth century, the Cottle County seat was a prosperous agricultural town in northwest Texas at the crossroads of two major highways, but by the year 2000, the yellow-brick 1930s-era Cottle County Courthouse was hemmed in on all sides by abandoned brick buildings, several of them collapsed in on themselves. When older residents looked at the once-sturdy structures, they remembered thriving cafés, drugstores, a couple of department stores, a variety store, hardware stores, a large hotel. They saw people on the sidewalks, traffic on the square.

Along red-brick residential streets a few blocks beyond downtown, empty houses blasted by sand and West Texas wind slowly crumbled. A population of more than 3,000 in the 1970s shrank to about 1,100 in 2010. The perennially tough Paducah Dragons football team switched from the eleven-player game to six players.

Paducah's decline had to do with boll weevils and a bad economy, nagging drought and the perils of dryland farming, as well as a government program that took farmland out of production. In 1980, Cottle County boasted more than seventy farmers; three decades later, only seven were still farming. Paducah had lost its core reason for being.[2]

Paducah native Rachel Richards, writing about her hometown in the September 1999 issue of *Texas Co-op Power*, recalled a saying her father occasionally repeated while enjoying a second cup of coffee at the breakfast table. "You know, Rachel," he would say, "in a small town we work people just the opposite of the way we work cattle. In a successful cattle-operation, we send the aging or poor performers off to the market, but with people we always manage to ship off the young and able." (W. Q. and Regina Richards saw their five daughters leave Paducah after college, returning only for visits.)[3]

Bartlett, proud birthplace of the nation's electric cooperatives, was another small Texas town felled by the boll weevil, soil depletion, and a bad economy. Its population peaked at 2,200 in 1914. Blackland farmers in Williamson County produced more cotton in 1900 than farmers in any other county in Texas. Thirty years later, the self-proclaimed "Buckle on the Cotton Belt" was sagging as farmers gave up and moved to the city.[4]

Like John Steinbeck's indomitable Joad family in *The Grapes of Wrath*, rural Texans were hitching rides, hopping freight trains, or loading up whatever ramshackle vehicle they might own and leaving behind the worn-out cotton farms, played-out logging camps, and, if they were black or Latino, small-town prejudice and oppression. In Texas and throughout the South, African Americans were among the first to leave, although the co-ops had little to say about what would come to be called the Great Migration. They saw opportunities in urban factories during World War I and in the 1920s, and like the Israelites they read about in their Bibles, they journeyed to a new land to find a better life, in their case to cities, particularly in the North. The Depression worsened the problems that had plagued southern agriculture throughout the 1920s, and when World War II came and defense plants called for workers, southerners, black and white, responded with a mass migration.

It wasn't just the small towns that were emptying out. In the second half of the twentieth century and beyond, total agricultural output in the United States more than doubled, while the number of self-employed and family farmworkers dropped by roughly 75 percent. By 2020, one farmer was doing the work of eight agricultural workers in the post–World War II generation. On the twenty-fifth anniversary of the REA, in 1960, Texas had ninety-six co-ops, more than any other state. Iowa was next with fifty-five. And yet demographers could see problems massing just over the horizon. In 1950, the rural population of Texas fell below 50 percent for the first time. Electric co-op leaders and the REA realized that the emptying-out of rural Texas was a threat to the co-op way of life.[5]

"Why do they leave the farm? Or to put it another way, why have so few veterans and ex-war workers returned to Texas farms since peace was declared?" *Texas Co-op Power* asked in March 1948. The newspaper turned to Joe R. Motheral, a rural-life economist at Texas A&M. Relying on a recent survey, conducted by the Bureau of Agricultural Economics of the US Department of Agriculture, in cooperation with the Texas A&M College

Agricultural Experiment Station, Motheral noted that only 1,773,000 Texans were living on farms in January 1947. Despite a high birth rate in relation to the death rate in rural areas, the state's postwar farm population was lagging far behind the prewar figure of 2,107,000 residents. Only three-fourths as many Texans were living on farms in 1948 as in 1930. Granted, it had been worse. During the war, farm population dropped to 1,577,000, the lowest since before the turn of the century. Recovery had been slow, Motheral noted, because of high postwar levels of urban employment, combined with drastic reductions in farm labor requirements thanks to mechanized farming, as well as changing crop and livestock methods.

TCP found it odd that farmers who responded to the survey tended to minimize such basic economic explanations. Many of them suggested that veterans weren't particularly eager to go into farming because of the "lack of city conveniences." They listed poor roads, lack of electricity, and generally unappealing living conditions on farms as the main reasons veterans weren't coming back to the land. Apparently, rural Texas was experiencing a WWII variation on the WWI ditty "How Ya Gonna Keep 'Em Down on the Farm after They've Seen Paree?"[6]

Young people were being lured from the farm to the city at a rate the US Census Bureau in 1950 called "alarming." According to a bureau survey, only eighty-seven people under age twenty-five were living on farms for every one hundred under twenty-five living on farms in 1940. The actual decline in farm population from 1940 to 1949 was 2,290,000, but of that decrease, 90 percent were under twenty-five years of age. The Census Bureau noted that the decline in farm population from 1940 to 1949 was all the more striking in view of the marked increase in the nation's population as a whole, which grew by about seventeen million during that period.

Texas Co-op Power noted that rural population decline was nothing new. It had been going on, with minor variations, for more than thirty years. In 1917, the nation's farms supported thirty-two million people. At the same time, agricultural experts pointed out that the increase in farm productivity and in the amount produced per worker had more than kept pace with the loss of laborers.

In the cities, meanwhile, jobs were plentiful as the nation revived from both the Depression and a wartime economy. The Census Bureau reported twelve million manufacturing jobs in 1947, rising to fourteen million jobs two years later. That number was in contrast to the 1937 level of eight million, which rose to fourteen million during the war because many women were involved in production work.

Young people were not only making more money in town after the war but were also able to buy houses at relatively affordable prices. "The family setting out today to buy a home has a wider choice, materials, designs and construction is better, and builders are offering excellent dollar values," *TCP* reported. "House costs have dropped as much as 25 percent, and the finished units are much more attractive and far better equipped than those available during the first year or two after the war. The house that sold for about $14,000 in 1946 and 1947 now can be purchased for somewhat over $10,000; and, instead of being only a shell, with no kitchen equipment included in the selling price, it is well equipped, even a TV set being built into the living-room in some instances."

Two years later, farm population was still in decline, according to the Bureau of Agricultural Economics. Total farm population was about the same as it had been fifty years earlier—about 23.5 million—while the nation's population as a whole had experienced an increase of about 75 million. The BAE pointed out that more ex-farmers were moving into small or midsize towns or suburban areas rather than big cities. "As for the youngsters," *Texas Co-op Power* reported, "from one-third to one-half of those born on farms now move to town. More girls than boys find attractive opportunities in town. As a result, many localities discover there aren't enough girls staying on farms to furnish wives for the farm boys."

The BAE report noted that both urban and rural birth rates had declined for many decades, reaching a low point during the Depression. They had been on the rise since then, but the farm birth rate was failing to keep pace with the birth rate in cities. Fifty years ago, BAE pointed out, farm women had 77 percent more children than urban women. In the 1950s, they were having about 51 percent more. The BAE also noted an increase in part-time farming. Many farmers had taken jobs in business or in factories, while many businesspeople and industrial workers were tending small truck farms or raising cattle on small farms. "The experts believe this situation is fostered by the increase in farm home conveniences and the shortening of the industrial work week," *TCP* noted.[7]

Why do they leave the farm? The answer became even more starkly obvious in 1950, when a suffocating drought set in that lasted until the spring of 1957. West Texas ranchers resorted to feeding cows and goats prickly pears to keep them alive. Farmers saw their crops wither and their soil blow away. Only 10 of the state's 254 counties escaped being declared "federal drought disaster" areas. Throughout the state, the day came when family members sat at the kitchen table and contemplated the unthinkable—packing up and leaving their homes in the country. The drought cost the state almost a hundred thousand family farms and ranches.[8]

"From a sociological point of view, [Agriculture] Department specialists say these population trends away from the farm are nothing to worry about," *TCP* noted. "In fact, they say it is a healthy situation that exists in practically every technically advanced country in the world. The young people are not all needed on the farm, so they go into fields where the best opportunities exist—medicine, teaching, mechanics, business and industry."

The sociologists may not have worried about the decline in the nation's rural population, but the co-ops realized the threat it represented. Abandoned farmhouses and hollowed-out small towns were tangible evidence of a dwindling number of co-op members. Michael E. Sloan, manager of the Cooke County EC (now PenTex Energy), told TEC's annual meeting in April 1961 that the future of electric co-ops, despite their successes, was by no means ensured because of rural decline. "Changes in the agricultural and rural economy over the past 15 or 20 years have eliminated many of our residential and farm consumers," he noted. "Schools and churches have disappeared from some of our communities, leaving them deserted and without identity. Some of the farmers still residing on small tracts of land today will be the last ones to live there. Much of the land in some communities has been rendered unproductive due to the lack of proper conservation practices."

Although Cooke and Montague Counties, Cooke County EC's original territory, had not suffered too badly, Sloan said, other co-ops in Texas and across the country were struggling. It was a bleak outlook, although Sloan assured his listeners the co-ops could survive and thrive if they recognized that their well-being was bound up with the economic prosperity of the region they served. "Our organization has grown into an impressive community institution, and this carries a responsibility on our part to assume a role as leader and take an active part in developing the small towns and rural areas that we serve," Sloan told his listeners. "We can no longer take the attitude that the sole objective of our cooperative is to deliver electricity. We must not continue to sit back and watch our local youngsters go off to the large cities looking for jobs, for the retention of our young people is the key to the future growth of our area." He continued, "We must realize that the electricity which we sell can be used to turn the wheels of rural industries, processing plants and other community enterprises that will provide the jobs necessary to retain the young people and to increase the standard of living for all residing in our area. This is also the key to the increased use of electricity, which we know is essential if your cooperative is to maintain cheap rates in the future."[9]

Later in the year, Richard M. Hausler, director of the REA's newly created Rural Areas Development Staff, echoed Sloan's concerns in an article published in *TCP*. Three out of four of America's rural young people left for the city before they were thirty, he wrote, because they couldn't find jobs in the small towns and rural areas and because they believed the cities offered them a better quality of life.

Co-op members, Hausler insisted, could help reverse mass migration by assisting their service areas in their efforts to develop good schools, including vocational training; up-to-date hospitals; good community facilities, including water and sewage; good housing and local financing; and a variety of career opportunities. Small towns, with the co-ops' assistance, had to reinvent themselves or die. "This is probably the greatest challenge that has ever faced rural America and the rural electric systems," Hausler wrote. "The potential is unlimited. The job will not be easy. But then, back when REA was getting started, who thought that rural people with technical assistance and credit from their government could build and successfully run a multi-million-dollar electrical system?"[10]

The tumultuous '60s careened toward a new decade. LBJ, the co-ops' stalwart friend in the White House, announced he would not run for re-election. Assassination claimed the lives of Dr. Martin Luther King Jr. and US senator Robert F. Kennedy Jr. Cities burned. The Vietnam War seemed endless. Co-op leaders were still expressing concern about "the rural-urban crisis," as Robert D. Partridge, newly appointed NRECA general manager, labeled the problem in May 1968. "It is called the 'rural-urban crisis,'" he wrote, "for good reason: It has its roots—and, I am convinced, its ultimate solution—in rural America. But its effects are no less devastating upon urban America."

The problem, in Partridge's view, was that millions of rural Americans had fled to the cities, and the cities were straining under the load of overpopulation. "With 70 percent of the people of this nation crowded into 1 percent of the land area, the cities are literally bursting at the seams," he wrote. "Yet hundreds of thousands of our rural people continue their trek—like moths drawn to a flame—into the cities every year."

Partridge suggested that slowing down and eventually reversing the Great Migration would not only make the nation's cities more manageable but also restore prosperity to rural areas, small towns, and villages. Two key elements, he suggested, were providing dependable electric service for the areas the co-ops served and "revitalizing the social and economic structures of our rural areas." Providing dependable electric service required adequate financing and power supply. He warned co-op members that the REA ran the risk of running out of money in 1969. He also contended that rural electrification had to have generation and transmission facilities of its own. Reversing the Great Migration also required revitalizing rural economies. "By working together on this vast undertaking, we can make the local rural electric system the focal point in communities across the nation, in this great effort to develop local industry and to improve local public facilities."[11]

The co-ops were still talking about reversing the great migration a decade later and beyond. For the most part, the task turned out to be too daunting—for co-ops and for other rural organizations, government entities, and individuals. Co-ops did their part to become an integral part of their communities, as their early leaders had encouraged. They backed economic development plans and got involved in volunteer endeavors and charitable efforts, and yet young people continued going away to college or the military and never coming back—unless, if they could afford to, they returned to raise children in what remained of the rural or small-town atmosphere they remembered.

As small towns and crossroads communities dwindled to almost nothing, the towns struggled to provide health care as hospitals consolidated. School districts were starved for students. County governments served fewer citizens and therefore provided fewer jobs. Census data indicated that about 40 percent of Americans lived on a farm in 1900; in 2020, the number was around 1 percent.

Not every little town succumbed to inevitable decline. In Jefferson, an old riverfront town in far northeast Texas, the ladies' garden club refused to allow graceful antebellum homes and historic downtown structures to be razed. The quiet, leafy town became home to more historical markers per capita than any town in the United States, and tourism kept the handsome community prosperous. Jefferson seemed to have trumped the legendary curse laid on it by industrialist Jay Gould.

Something similar happened in Fredericksburg (home to Central Texas EC), where the locals' respect for their tidy town's German heritage and handsome native-stone architecture made the Gillespie County seat one of the most popular tourist draws in the state. La Grange, Round Top, and Fayetteville, all in Fayette County, also capitalized on their quaintness and appreciation for historic preservation. So did Shiner, the self-proclaimed "Cleanest Little City in Texas," although a local brewery with a nationwide reputation also drew visitors and helped keep the town prosperous. Albany and Canadian relied on wealthy families who didn't forget their hometown when they made their fortunes elsewhere.

Those pleasant communities were exceptions to a disturbing rule, however. As the twentieth century marched toward the twenty-first, most small Texas towns were struggling. Texans had moved to town. They still wore cowboy hats and boots, but their southwestern attire most likely complemented pinstripe suits. They drove pickups, just like their rural forebears, but, with country music blaring over satellite radio, they likely were plying multilane freeways, not farm-to-market roads. They lived not on ranches but in suburban ranch-style houses. They worked in office towers, not on that fabled prairie where the deer and the antelope play.

In the early years of the new century—once the foreboding Y2K passed with no consequence—Texas was the second-largest state in terms of rural square miles, but population decline seemed inexorable. Even as Texas between 2000 and 2010 became the fastest-growing state in the union—claiming six of the largest twenty-five cities in the nation—small towns and rural areas continued to hollow out. The US Census Bureau reported that during the first decade of the twenty-first century, 79 of the state's 254 counties had declining populations. From 2010 to 2014, the number of counties in decline had risen to 102.

"These are long-term trends that are endemic and sort of built into the very structure of agriculture, which is designed to grow and scale and replace labor with capital," Matthew Sanderson, a former sociology and anthropology professor at Kansas State University and editor in chief of *Agriculture and Human Values*, said in a 2021 interview with *Rural Messenger*. Sanderson noted that the non-Hispanic white population in rural areas had been declining for decades, while the only growth was among the foreign-born immigrant population. Meanwhile, Texas farmers were experiencing a labor shortage. To cope, they turned to buying larger and more automated equipment, Pat McDowell, a Shamrock-area farmer and a Texas Farm Bureau director, told *Rural Messenger*. "And as fewer people work on farms, the effect flows downhill into rural communities.

With fewer residents purchasing from local stores and businesses, those companies shut their doors, forcing folks to travel to regional hub cities for goods and services," *Rural Messenger* noted.[12]

Political influence also waned. With rural areas attached to more populous urban and suburban districts, issues of concern to country people were an afterthought. Angela Arthur, a resident of Ralls in Crosby County, lived only thirty minutes west of Lubbock, but she told lawmakers at a 2020 Texas House redistricting committee hearing that her town was struggling, even as Lubbock prospered. "We have no major businesses outside of the local co-op," Arthur said. "Our churches, schools and hospitals just struggle to stay open."[13]

The Crosby County population dropped more than 15 percent between 2010 and 2020. In 1930, Crosby was home to 11,023 residents; in 1960, the population had declined slightly, to 10,347. In 2020, fewer than 6,000 people called the county home. Crosby was one of several counties losing population in the South Plains and Panhandle regions, traditional farming and ranching areas distant from cities. In addition to Crosby, the counties emptying out decade after decade included Dickens, Hale, Schleicher, and Floyd. All were served by electric co-ops.

"No one who lives in Dallas or Tarrant County is concerned about rural broadband access, because they have some of the best, fastest internet options available," Laramie Adams, a Texas Farm Bureau national political director, pointed out to the committee. "They probably don't contact their legislators to support rural broadband bills. But those in the Panhandle do, because it's a dire need in their daily lives."[14]

Texas was gradually becoming two states, sociologist Robert Cushing noted. "Where to build the next school can be an issue at one extreme. At the other, the question may be what is the next school to close," he wrote in the *Daily Yonder* in 2021. "Growing counties need to tax and invest in infrastructure, whereas the declining counties are losing the population base needed to sustain the existing infrastructure."[15]

A few co-ops went out of business over the years and others consolidated as rural Texas declined, but, paradoxically, a number of them found to their surprise that they didn't have to worry all that much about luring city folks to the country. The city came to them. Urban Texas arrived in the form of housing developments and subdivisions spreading like wildflowers across what only the day before had been rangeland or cotton fields. Some of the fastest-growing cities in the nation were spawning populous suburbs.

Small, snoozing towns were either engulfed by a spreading city or had become urban dynamos themselves.

Sara June Jo-Saebo, a midwesterner who moved to the Dallas area in the early 2000s, observed that

> the flat pastureland retreats from the city like a hotcake on a griddle. . . . You can look at Dallas on the computer through something called "satellite view" and you can see just how big it is and how the smaller cities are encroaching on each other. Some landowners are holding onto their small ranches inside these urban areas, made apparent on satellite view with squares of pasture that stand out like patched knees against the tangle of winding neighborhood streets connecting millions of households crowned with asphalt. There is no end to the asphalt shingles in Dallas, Texas.[16]

In Hutto, a small farming community twenty miles northeast of Austin, most everyone who gathered on fall Friday nights to cheer the Fighting Hippos through the years knew everyone else; the total population in 1990 was only 627. Twenty years later, the Williamson County town (served by Bartlett EC) was home to nearly 15,000 people. A decade later, the population had nearly doubled, to 28,000 people.

Bartlett itself, Hutto's neighbor, had been moldering away for decades. The farm town had tried various schemes over the years to stay alive, including hiring itself out to Hollywood in the 1990s. For a while it worked. In 1994, the town's brick streets and picturesque downtown buildings, most of them empty, served as a readymade set for *The Stars Fell on Henrietta*, a movie about the Texas oil boom in 1935, produced by Clint Eastwood's company. In 1998, Austin filmmaker Richard Linklater came to town to make segments of *The Newton Boys*, starring then–teenage heartthrob Matthew McConaughey. Two other productions, *The Whole Wide World*, a biopic of Cross Plains native and *Conan the Barbarian* creator Robert Howard, starring Renée Zellweger, and the TV series *Fear the Walking Dead*, also brought Bartlett back to life.

Alas, Hollywood was fickle, and Bartlett couldn't sustain the interest. City fathers lured a state prison to the area, but it closed after a while. As with Hutto, Austin's explosive growth brought Bartlett back to life permanently, beginning in the 2020s.

Austin fueled dramatic growth to the west, as well. Hill Country towns in the Pedernales EC service area saw some of the most dramatic gains not only in the state but also in the country. The population of Dripping Springs, ten miles west of Austin, increased by an astounding 316.89 percent between 2010 and 2020. With a 2020 population of 7,454, the "Gateway to the Hill Country" was projected to be home to 83,000 by 2023.

Blanco, a town on the picturesque Blanco River, was growing, as well. "We're right in the middle of all that," Blanco County commissioner Tommy Weir told the *Houston Chronicle* in 2018. "We've been mainly farming and hunting, so we don't have a commercial district or a hotel tax that would help us pay for all the trucks tearing up the roads or EMS costs that have doubled."[17]

It was the same story for once-small towns within the force field of the Metroplex. McKinney, northeast of Dallas, had a population of 15,193 in 1970, small enough for everyone in town to conceivably know Benji, the fictional canine who starred in a movie of the same name, set in McKinney. The town's population soared to 195,368 in 2020, making it the fastest-growing city in the nation. Benji would have to be much more wary of the traffic on his morning jaunts downtown.

Co-ops were being transformed. Pedernales and Central Texas in the Hill Country; Tri-County northwest of Fort Worth; Bluebonnet east of Austin; CoServ, Grayson-Collin, and United Cooperative Services in North Texas; and Farmers in northeast Texas all were scrambling to stay abreast of dizzying growth. Theirs was a much more desirable problem to have than the alternative.

The transition from rural to suburban wasn't always easy for the co-ops. The Tri-County EC experience was typical of several co-ops in fast-growing areas. By the early 1970s, because high gas prices had resulted in high rates from Brazos Electric Co-op, most of the cities in Tarrant County that had significant numbers of Tri-County members had asked, indeed demanded, that the co-op sell out and pull out. In most cases, the co-op held its ground, stipulating that a two-thirds vote of all its members would be required to approve such a move and noting that arbitrarily pulling someone's power would deprive them of property without due process. The co-op also joined calls for the creation of a Public Utility Commission, which would eventually rule on territorial rights for electric utilities.[18]

On June 18, 1976, disgruntled Tri-County members—most of them from the fast-growing northern reaches of Tarrant County—called a petition-driven membership meeting to be held at Kangaroo Stadium in Weatherford. The members who called the meeting wanted out of the co-op and resolved to move that the board be dismissed, the co-op dissolved, and all its assets liquidated. The motion never came to a vote and the co-op's board remained in place.

In 1977, after Azle city officials began actively campaigning for citizens to drop the co-op and move their power to Texas Electric, Tri-County sold its system in Azle to the Texas Electric Service Company for $250,000. The co-op regained some membership within the city as it expanded.

In 2021, Tri-County revised its bylaws to increase board membership from eight to nine—four from "urban" districts (north Tarrant County), four from "suburban" districts (Parker, Hood, Palo Pinto, Denton, Wise, and Jack Counties), and one from its "rural" district (the nine counties that make up the former B-K Co-op around Seymour).

Tri-County continued to grow. With more than one hundred thousand members in 2022, more than half of its membership was in Fort Worth, Keller, Southlake, Westlake, Colleyville, and other fast-growing cities that had overwhelmed fields, pastureland, and small towns.[19]

"Folks are still moving from the country to the city," Mike Williams, TEC's president and CEO, noted in 2021. "Many co-ops are, however, benefiting from the growth of the cities into the rural areas."[20]

The Texas legislature helped, however inadvertently. In 1975, lawmakers passed landmark legislation that locked in certified service territories, which meant that suburban growth redounded to the benefit of electric co-ops. Spite lines became distant memories.

"WE HAVE TO GET LARGER"

Despite the ongoing concern about Texans leaving small towns and family farms in the rearview mirror, a U-Haul trailer hitched to their vehicle, the state's electric co-ops in the 1960s were feeling ambitious. They had friends in the White House in Presidents Kennedy and Johnson, as well as congressional support, and in Austin, rural Texans still controlled state government. The co-ops resolved to liberate themselves from their dependence on their investor-owned neighbors for wholesale power. They were well aware that as long as they had to depend on the private power companies, they were at their mercy. They never felt comfortable relying on businesses that had tried to force them out of existence.

The co-ops were too small to generate and transmit electricity on their own, but they had a solution. Back in 1937, Texas legislation modeled on the Electric Cooperative Corporation Act and the federal rules governing the REA gave individual co-ops the authority to band together to form so-called generation and transmission cooperatives. The G&Ts, essentially cooperating cooperatives, would come together to obtain REA financing to build their own systems to meet the wholesale power needs of distribution co-op members. The G&Ts, in essence, offered economies of scale, allowing cooperatives to build power plants and the expensive high-voltage transmission systems they had to have to deliver power to their members.

According to REA rules, loans could be made for generation and transmission when an adequate source was unavailable, when private sources demanded unreasonable conditions or limitations, when wholesale rates were unreasonably high, or when savings were possible. Most co-ops could easily meet at least one of those criteria and thus would be able to break the chains that bound them to competitors that had perennially wished them ill.

Often, the co-ops found themselves burdened with dual rates: a regular rate for rural residential customers and a much higher rate for coveted industrial users. It was an old tactic: the IOUs were trying to force the co-ops into the less lucrative markets while keeping the most profitable for themselves.

In 1961, REA administrator Norman M. Clapp added another criterion for creating G&Ts. It came to be called "the security criterion." Speaking to the annual meeting of a G&T system called Western Farmers Electric Cooperative in Anadarko, Oklahoma, Clapp noted that adequacy, dependability, and lower cost were not enough. "We also must be sure the cooperatives enjoy a supply of power—now and in the years ahead—that will guarantee the cooperative device a permanent place in the American power industry," he said. "One way you can obtain this guarantee is through power supply contracts that are fair and negotiated in good faith. The other way is by generating your own power, as you people here are doing. We intend to use our generation transmission authority when our borrowers are unable to obtain the security they need through power contracts."[1]

As far back as 1945, investor-owned utilities had battled the idea of G&Ts. Congressman Bob Poage of Waco and NRECA's Clyde Ellis warned of a gigantic well-organized campaign to sabotage the co-ops by cutting off sources of wholesale power. They charged that the IOUs were looking for lawmakers in both Washington and Austin who favored restrictions that would prevent the REA from lending money to cooperatives to build generating plants and transmission lines, while attempting at the same time to sidetrack congressional appropriations for public power and flood control projects. "With these two sources of wholesale power eliminated," *Texas Co-op Power* warned, "the co-ops would be completely at the mercy of the private utility companies in the purchase of wholesale energy."[2]

"We are fast approaching the most critical period in the history of the REA program," Poage told co-op leaders from Texas, New Mexico, and Arizona in a November meeting in Austin. Poage at the time was sponsor of a "full-coverage" bill in the House that would make available $500 million to the REA for loans to co-ops to carry on a three-year program of rural power line construction. "There is no force in our economic structure so effective as competition when it comes to securing fair prices and good services," the congressman told the Austin gathering. "For example, in

Central Texas the Brazos Transmission Cooperative has been able to offer private companies effective competition. As a result, the private companies reduced their wholesale rates to cooperatives by 60 percent during a six-month period. It all boils down simply to a question of whether or not we believe that the government should help farmers get credit with which to put their business on a sound footing."

The Waco congressman expressed his belief that a government that extended credit to railroads and banks through the Reconstruction Finance Corporation could well afford to extend credit to farmer-owned cooperatives at an interest rate that had actually earned a substantial profit for the government. Poage's bill got bottled up in House committees in early 1946, leaving the REA only $200 million in loan money until June 1947. "Don't ever give up your right to generate your own power," the congressman urged the co-ops.[3]

A dozen years later, skirmishing continued. Generation and transmission cooperatives "have stirred up the monopoly utility companies mightily," *Texas Co-op Power* reported in September 1962. "And why do the utility companies get so worked up about generation and transmission? . . . It boils down to the fact that the commercial power companies are losing their long-established and effective control over the co-ops."

In an article for the September 1962 edition of *Texas Co-op Power*, NRECA correspondent Dick Wilson noted that rural electric co-ops generated only 16 percent of their own wholesale power supply, while buying 39 percent from federal sources, 38 percent from commercial power companies, and about 6 percent from public power systems. The for-profit utilities had little to worry about, the correspondent noted. The cooperatives were nowhere near taking over the electric power business.

The problem the co-ops faced, Wilson pointed out, was that they were finding it more and more difficult to help their rural members obtain a reliable supply of electric power at the lowest possible cost. Power shortages and the high rates the commercial power companies charged for wholesale power were the twin culprits. The only solution was for the co-ops to build their own power plants and transmit electricity to their member-consumers.[4]

Concerned about the reliability and cost of wholesale power, eleven Central Texas distribution co-ops had organized the state's first generation and transmission co-op in 1941. Headquartered in Waco (in Bob Poage's congressional district), it was called the Brazos River Transmission Electric Cooperative, Inc., later shortened to Brazos Electric Power Cooperative.

The name, of course, recognized the storied river the Spaniards christened Río de los Brazos de Dios.

The river, 1,300 miles in length, is quintessential Texas. The first Anglo-American settlements in Texas had sprung up beside its banks. The state's first capital was at San Felipe de Austin, on the banks of the Brazos. The document declaring Texas independence was drafted and signed at Washington-on-the-Brazos, a rude settlement a few miles north of San Felipe. The co-op's headquarters in downtown Waco was a few blocks from the river.

The river that provided the co-op its name began providing the power that would nourish the new co-op the same year it was organized, but only after a long, bitter fight. "The private power companies had fought the project, as they always oppose all efforts to control rivers where such control involves the production of hydro-electric power which might provide competition to them," *TCP* editor George Haggard noted. It took six years, but the co-ops prevailed. "It now appears that the people are going to be able to use their own power without paying tribute to the utility companies in the process," Haggard wrote. In 1941, Brazos built its first transmission line to generate electricity at Possum Kingdom Lake in North Texas. Brazos distributed the power produced at Possum Kingdom, free of charge, to eighteen co-ops in the Brazos River watershed. "These co-ops," Haggard noted, "are owned and controlled by 30,000 Texas farmers."[5]

The synergistic partnership between the Brazos River Authority and Brazos Electric Power Cooperative to provide power to rural Texans was highlighted in a March 26 editorial from the *Marlin Democrat*, the weekly newspaper that served the Falls County seat: "The completion of the Possum Kingdom Dam marked the first step in the realization of a dream of many years. . . . The development of hydro-electric power is a by-product of the major aim of conservation and reclamation of the rich farm lands which have been denuded by unchecked floods. The sale of power is on a non-profit basis. The aim is to reach rural areas not served by the existing utilities, thus giving the farms the advantage of electric power and light and rendering rural life more attractive and comfortable."[6]

Forty years later, Brazos EC was the largest G&T cooperative in the state, selling wholesale power to twenty member co-ops serving 229,000 consumers in sixty-six counties along the Brazos River. As the co-op noted in its 1986 annual report, farmers and ranchers were hardly the only consumers of electricity in the co-op's service area. The co-op noted that "modern, multi-million-dollar companies such as the Westinghouse Electronic Assembly Plant in College Station and Hexcel Corporation in Graham thrive in towns along the Brazos River, providing jobs, prosperity, and most importantly, products to carry our world into the 21st century."[7]

Eventually, nine G&T cooperatives were organized. In addition to Brazos, they were Western Farmers EC (1941), based in Anadarko, Oklahoma; Nursery-based South Texas EC (1944); Longview-based Northeast Texas EC (1972); Jourdanton-based San Miguel EC (1977); Rockwall-based Rayburn EC (1979); Amarillo-based Golden Spread EC (1984); and Nacogdoches-based East Texas EC (1987). The Lower Colorado River Authority was also considered a G&T.

The story of a G&T named in honor of Sam Rayburn was typical. Rayburn Country was organized in 1979 to serve seven distribution co-ops founded in the mid-1930s by farmers growing cotton and raising dairy cattle. All seven of the original member co-ops, located in an arc from southeast to northeast of Dallas, received most of their electric power from Texas Power & Light. The co-ops were constantly dealing with escalating power costs and never-ending feuds with TP&L, Texas Electric Service Company (Fort Worth), and Dallas Power & Light Company over wholesale power costs. (The three utilities merged in the early 1980s to form TXU Energy).[8]

As Ray Raymond, a longtime general manager of Kaufman County EC and one of the Rayburn Country founders, explained in 2019, TP&L would impose fees and charges in order to maintain retail rates about two cents below the rates they charged their wholesale cooperative customers. "That caused us a lot of headaches," he recalled, "with people complaining about the cost of our power. I was always saying that I was going to do something that would make it possible for the cooperatives to be cheaper than TP&L."[9]

Raymond lived up to the promise he had made to himself. In conjunction with co-ops in Louisiana, the Texas co-ops that were part of Rayburn Country were able to compete from a position of strength with both TP&L and the Southwestern Power Administration (SWPA).[10]

On June 27, 1979, Earnest Casstevens, an Austin lawyer serving as a consultant to the seven co-ops, filed the official charter with the secretary of state's office. Rayburn—the name for the cooperative was a unanimous choice—was incorporated as a nonprofit company, pursuant to the Texas Electric Cooperative Corporation Act. Raymond, still acting as general manager of Kaufman County EC at the time, would serve as president of the new co-op and as chair of the Rayburn board.[11]

Raymond had spoken for years about the importance of uniting co-ops to gain bargaining power in wholesale purchased power negotiations. “That was my main focus from the very beginning,” he said. “I said we have to get larger.”[12]

"THE FOLKS SENT A MESSAGE, AND THE LAWMAKERS RESPONDED"

In June 1975, Texas became the last state in the union to establish a commission to oversee and regulate public utilities. The Texas Public Utility Regulatory Act, long sought by consumer advocates and the state's electric cooperatives, passed the sixty-fourth session of the Texas legislature in part because of a suicide note. The tragic missive, penned a few months earlier by a Southwestern Bell executive in charge of the company's Texas operations, read as follows: "Watergate is a gnat compared to the Bell Telephone System."[1]

The executive, T. O. Gravitt, wasn't bluffing. After his death by carbon monoxide poisoning at his Dallas home, his family and a former associate, James Ashley, filed a $26 million lawsuit that kept Ma Bell in the news for months, besmirching her reputation and costing her real money.

The plaintiffs alleged the phone company was guilty of "corporate rapacities," harassment, bugging, improperly influencing public officials, using corporate funds for political contributions, and overcharging customers. They charged that Bell misled local rate regulators into approving higher-than-necessary rates, used Bell system buying power to influence city council members who voted on rate increases, and maintained a political slush fund collected from senior executives who allegedly received raises to cover campaign contributions.

Bell responded by charging that the company had been investigating Gravitt and Ashley for financial irregularities and for pressuring female employees into sex in exchange for promotions. During the trial, Bell produced about a dozen of its female employees who testified that one or the other of the executives had either propositioned them or had sex with them, or that they had witnessed others doing so.

Pat Maloney, the colorful San Antonio attorney representing the plaintiffs, said testimony had demonstrated widespread promiscuity among other Bell executives. "Bell will never again have its old, prestigious conservative image," he contended. Whatever the truth of the allegations, the scandal was an expensive ordeal for Bell. The company lost millions of dollars in rate-increase requests throughout the five-state Southwestern Bell territory. The jury eventually awarded Gravitt's widow and Ashley $3 million, much less than the $26 million the plaintiffs had requested, although intangible costs were much higher.[2]

The daily drumbeat of alleged chicanery fueled public disgust and cynicism, not just against Southwestern Bell but also against elected officials and overbearing public utility companies. Negative opinions were already rampant in the wake of Watergate and President Richard Nixon's recent resignation.

Watergate had broken open not long after a Texas banking and bribery scandal called Sharpstown destroyed the political careers of high-level elected officials, including the House speaker, the governor, and a young lieutenant governor destined for the White House (according to Lyndon Johnson). Sharpstown prompted not only a general housecleaning at the capitol but also a raft of legislation designed to strengthen public accountability and transparency. Creating a Public Utility Commission was a continuation of the same reformist impulse.

Texas lawmakers had attempted to regulate public utility companies—telephone, electric, water, and sewer—since the early years of the twentieth century but could never agree on a mechanism. They increased the power of local authorities occasionally through the years, yet city officials were still constrained. Some lawmakers proposed placing utilities under the regulation of already existing agencies—the Texas Railroad Commission, for example—but nothing ever came of the proposals. Legislative efforts to create a new commission cropped up in 1913, 1929, 1946, 1957, 1969, and 1973, but the power companies and their legislative allies beat back every effort.

"The climate is right for a public utility commission," said Latham Boone, D-Navasota, one of the co-sponsors of HB 819, the Texas Public Utility Regulatory Act of 1975. "With the recent disclosure of alleged improprieties on the part of Southwestern Bell Telephone, the high-rise utility rates being felt across the state and the awareness of the citizens of Texas that our state is the only one that does not provide statewide regulation of any of the utilities, I think we'll have the Texas Public Utility Commission."[3]

In the opposite wing of the capitol, state senator Ron Clower, D-Garland, called one of the Bell memos on rate-making "the most shocking thing I have ever read." Clower, chair of a Senate subcommittee on consumer affairs, sponsored a bill that would set up a three-person elected public utilities commission with authority to regulate phone service, electric utilities, natural gas, and private water and sewage companies. Municipally owned utilities would be exempt, and cities could retain local control over rates by majority vote.

Governor Dolph Briscoe preferred a state corps of rate and regulatory experts rather than a commission, the *Texas Observer* reported in December 1974. Under the governor's plan, cities could call on the experts when they wanted help, while unincorporated areas would have to rely on the experts. Lieutenant Governor Bill Hobby seemed to look with favor on Clower's bill and had endorsed a utilities commission with elected members. House speaker Bill Clayton had long opposed utilities regulation, "but he seems to be softening," the *Observer* reported.

State representative Ed Watson was another HB 819 cosponsor. The Deer Park Democrat warned that the utility companies "will do everything in their power to kill this legislation."[4]

The companies fought, indeed, but they lost. House Bill 819, sponsored by state representative John Wilson (D-La Grange), passed in the waning days of the session by a vote of 22–9 in the Senate and 120–24 in the House. As Boone noted, Wilson's bill was a legislative response to public disgust with not only political scandal but also ever-higher utility bills. For months, electric co-ops and the big utility companies both tried to respond to customer and member complaints about what they were paying. Co-op officials held meetings among themselves and tried to craft an explanation, one that pointed out that the utilities were at the mercy of the natural-gas companies.

The public was in no mood for explanations. They wanted a solution. Texans complained to their elected officials. They organized and protested. They showed up at the capitol, sent letters and telegrams vowing to vote their representatives out of office if they didn't respond. They were trying to make enough noise to drown out the powerful utilities, whose lobbyists were insisting that the rate problem would solve itself without the creation of a new regulatory agency.

Co-op members of a certain age no doubt remembered the infuriating stubbornness they had encountered when they had approached the utilities more than a half century earlier, requesting service. This time, though, the wealthy and powerful could not rely on spite lines and other devious blandishments.

"The timing was right for democracy at work," *Texas Co-op Power* proclaimed. "Simply put, the people of Texas sent a forceful message to Austin: 'We want a utilities commission! It might help.' And the legislators responded—if not too eagerly."[5]

Walter Richter, a wry and knowing former state senator and a populist ally of the co-ops, echoed *TCP*'s editorial. "Those lopsided votes, incidentally," Richter wrote in "Poor Richter's Almanac," his regular *TCP* column, "reflect once again the power of the people in the democratic process, because I would wager that roughly half and maybe even more of the legislators were personally philosophically opposed to Texas having a commission. The folks sent a message, and the lawmakers responded."

Richter, director of government relations for Texas Electric Cooperatives from 1972 until his retirement in 1982, waxed philosophical: "Incidentally, the pro-commission people in my opinion did not place sufficient emphasis on the completely irrefutable argument that any industry which is by nature monopolistic is also by nature selfish and therefore must be regulated in some manner if the public interest is to be accommodated."[6]

Governor Dolph Briscoe appointed three commissioners—Garrett Morris of Fort Worth, Alan R. Erwin of Baytown, and George M. Cowden of Dallas. They would serve staggered six-year terms.

The *Fort Worth Star-Telegram* described Morris, the board's first chair, as "a political bridesmaid, a quiet intense man who has run dozens of campaigns for others but who has never sought a political office for himself." The *Austin American-Statesman* described him as "a Gary Cooper in a narrow-brimmed hat," a description Morris said he liked.

Erwin was only thirty when the governor named him to the board. A former journalist and later a public relations executive, he had written speeches for Briscoe. After his Public Utility Commission (PUC) service, he wrote a novel about the experience.

Cowden was a graduate of Baylor University Law School and a Dallas attorney who had served two terms in the Texas House representing the Waco area. He had also served as chair of the State Board of Insurance. He told *TCP* he enjoyed hunting and whittling and considered himself a "good ol' boy from Pearsall." He said that twenty of his wooden caricatures of US presidents were on display at the Presidential Museum in Odessa. (Presumably, the ability to carve elected officials down to size would be a valuable asset in his new job.)

Cowden and colleagues had three initial chores: implement reasonable utility rates, ensure adequate utility service for the people of Texas, and provide a fair profit for the utilities that would allow them to expand and attract loans to build expensive new power plants. The new commission also would address environmental issues and oversee intrastate telephone rates and services, while leaving utilities in incorporated areas to city oversight. On September 1, 1976, the PUC assumed authority over rates.

"Even with the new utility commission, any reduction in electric rates will be hard to come by," Frank Youngblood of the Texas Railroad Commission told a gathering of electric co-op leaders. "Our generation fuels simply are in short supply and high demand. That means high prices."

For the co-ops, the most important provision by far, and a prerequisite for their support, was area certification. Certification meant that each service area of the state would be assigned to a utility, presumably the one already serving the area. No other utility would be allowed to expand into the assigned area of another. If the PUC deemed the service inadequate, another utility would be designated to serve.

Vital to a fast-growing, urbanizing state, certification applied to annexed or incorporated areas. An electric co-op serving an area just outside a voracious Dallas or Fort Worth—Tri-County, for example—would be able to continue serving its present members and extend service to new members, even if the area served was annexed by the expanding city. No longer would a utility be able to invade the territory of another. No longer would cities be allowed to take over a cooperative's service area simply by annexing it.

"Financiers will feel secure that a cooperative will always have enough revenues in its certified area to repay debt," Randall Mallory wrote in *Texas Co-op Power*. "Cooperatives will be able to secure the vast amounts of money they will need to meet the future demands of their member-consumers."[7]

Area certification was "a major breakthrough," Richter wrote. "This will be difficult to implement," he added, "and will almost certainly involve some lawsuits where distribution lines currently overlap, but once established, it should be a boon to everyone, notably the consumer who has been paying for a great deal of costly, wasteful duplication of facilities."[8]

Stamford EC co-op manager Charles Stenholm, a future West Texas congressman, was hopeful. "In Texas," he told an impromptu gathering of attorneys and co-op managers at TEC's annual meeting in Amarillo, "we in rural areas are the future. The new commission secures that future and gives us the vehicle with which—with continued good service by our cooperatives—we can meet the challenges of the future for rural Texas."

"It's Certified!" was the theme of the Amarillo meeting in August and the topic of conversation among attendees, formally and informally. A few months later, the co-ops were working on implementing the new regime. The experience of the Kaufman County EC was typical.[9]

The first step, as Neal Johnson and Ray Raymond explained in their 1998 history of Kaufman County EC, was to apply to the PUC for a certificate of public convenience and necessity. That was the procedure that would determine the service territory boundaries for the utilities. Every utility sent applications and maps of their respective service territories to the PUC in early 1976, and the commission was required by law to act on the applications by the end of August. "Certification worked fairly well," Raymond recalled. The co-op management team and Texas Power & Light spent hours at a time hashing out where the lines would be. On August 4, co-op board members met at 7:00 a.m. to review maps showing the certified territories for exclusive and joint certification. The board voted that day to submit the maps to the PUC.

One of the first tests of the new law was a question raised by the PUC staff about an issue that had roiled co-ops almost from their beginning: Does the Public Utility Regulatory Act (PURA) allow electric co-ops to serve inside a city's limits?[10]

Since 1955, when Judge Jack Roberts had ruled against Upshur Rural EC's effort to provide service inside the city limits of Gilmer, co-ops had lost growth territory as cities expanded. PUC attorney Martha Terry raised the sticky issue, and the power companies responded with one voice. No, they insisted. PURA did not change the limitations that had hemmed in co-ops for two decades. Co-op attorneys contended that PURA abolished the limitations.

At the end of a day-long hearing, newly installed chair Morris ruled in favor of the co-ops. They would be allowed to serve in the incorporated areas of cities, because the PUC had authority to certify co-ops to serve within city limits.

Raymond, Kaufman EC's board president at the time of the ruling, breathed easy. "When cities began to incorporate and expand in our service area, I felt good that we were certified to those areas and that we could remain and serve the growth that occurred," he said.[11]

Raymond wasn't the only one relieved by the ruling. In the coming years, every electric co-op within shouting distance of the state's astronomically growing cities had reason to be grateful to a Southwestern Bell scandal, to members and customers fighting mad about high electric rates, to lawmakers responsive to the will of the people, and to a Public Utility Commission that, on this particular issue, ruled against the big power companies. The PUC wouldn't always side with the co-ops in the years to come, but the early certification ruling kept them in the game.

"WE HAVEN'T BEEN VERY GOOD AT REACHING OUT"

In the August 1972 issue of *Texas Co-op Power*, writer Randall S. Mallory profiled an eighteen-year-old soon-to-be Texas A&M freshman named Frederick McClure. Eleven months earlier, the young East Texan had been elected state president of the Future Farmers of America and had spent much of his tenure in office driving around the state and "selling" FFA in small towns and big cities. He also was visiting the state's 450 local FFA chapters and many of its fifty-one thousand members. "Like a catfish in fresh water, Fred McClure thrives on opportunities to meet new people and different experiences in and outside of FFA," Mallory wrote. "'On the road, I just groove out on the scenery and new faces I see, and sing along with the radio a lot,' Fred says."

His parents, F. D. and Mayme McClure of San Augustine, were members of Deep East Texas Electric Cooperative. Mr. McClure was principal of the San Augustine Intermediate School, while Mrs. McClure was a longtime school-district nurse. They had ample reason to be proud of their son.

Salutatorian of his graduating class, he had served as president of the Greenhand FFA Chapter and had participated in public-speaking contests and on forestry and meat-judging teams. Later, he was president of the San Augustine FFA Chapter, and in succession, the Piney Woods District FFA Association and the Area 9 FFA Association. All were steps along the way to the state presidency. He told Mallory he hoped to receive the American FFA Degree, awarded to only 1 percent of the nation's Future Farmers.

Young McClure was an accomplished organist and singer and had served twice as accompanist for the National FFA Chorus and twice as soloist. He also served as drum major for the National FFA Band. Listed in *Who's Who among American High School Students*, he planned to use several scholarships he had received to major in either agricultural economics or biochemistry at Texas A&M.[1]

Texas Co-op Power's profile of the impressive young man was notable for one other reason: Fred McClure was the first African American to be featured in the publication's twenty-eight-year history. Somehow, *Texas Co-op Power* had managed to publish more than three hundred monthly issues, from 1944 until 1972, without acknowledging that African Americans made up more than 12 percent of the population of a state that was home to more African Americans than any other. Fred McClure's parents weren't the only African American electric co-op members in Texas, but if *TCP* was any guide, they were invisible (as, to a slightly lesser extent, was the fastest-growing demographic group in Texas, Hispanics or Latin Americans).

Issue after issue of the publication offered up group photos of board members, invariably white men middle aged and older. General managers were white men, middle aged and older. Beauty contestants, a *TCP* staple for several years, were young white women. Young white men and women won FFA and 4-H honors. White families, never black families, showed off their new all-electric homes and their electric time-and energy-saving farm devices. White women shared their recipes, never black women (despite an incredibly rich heritage of southern cooking). Crowds that gathered for annual meetings in auditoriums and on high school football fields were all white (except for Hispanic members at Magic Valley and other South Texas co-ops). Until the 1980s, no black co-op members had ever served as board members. None served as co-op executives.

Readers of *Texas Co-op Power* could only assume that black co-op members, if they existed at all, were uninterested and uninvolved in serious issues that came up regularly in the magazine—rural housing, education, transportation, energy savings, rural medical care, water issues, the environment, electric rates, the future of the REA, farm policy, state and national elections. Only white women, it seemed, were interested in dress patterns, recipes, and the latest kitchen appliances. Illustrations and photos accompanying the publication's advertising suggested that only whites might want to purchase life insurance, modular homes, "oriental health sandals," choir robes, wigs, weight-reduction pills, dwarf-sized trees, vinyl siding, flowering cherry hedges, custom-made bowie knives, or finely detailed classic car models.

Managing to ignore such a significant and influential percentage of the population for almost three decades was astounding, almost a tour de force, but it was a reflection of society at large in a southern state, where public schools were segregated into the 1960s (despite the 1954 *Brown v. Board of Education* ruling), and where black athletes did not play for Southwest Conference teams until Texas Longhorns head football coach Darrell Royal in the '60s got tired of losing to teams from the Midwest and West that recruited black players from Texas. *TCP* catered to a Texas where neighborhoods remained segregated long after the Civil Rights Act of 1964, where black Texans went to their churches and whites to theirs. Rural Texas was as racially segregated as urban and suburban Texas, if not more so. *TCP* also was a reflection of electric co-ops, in Texas and elsewhere. "We haven't been very good at reaching out," TEC president Mike Williams acknowledged years later.[2]

Failure to reach out was not solely a Texas problem. "Since their formation, many RECs have either embraced or fallen prey to old boys' networks of board members—older white men, by and large—who have let any democratic process that did exist erode over time," Kate Aranoff, writing in *Dissent* magazine in 2017, observed. Aranoff noted that among NRECA's forty-seven-member board of directors in 2017, all but two were white.[3]

A 2016 study conducted by an organization called the Rural Power Project found that electric cooperatives in the American South, including Texas, were not living up to their founding ideals. African Americans, Hispanics, and women were vastly underrepresented in positions of authority.

"Transparency is rare and too many rules and procedures are designed to maintain a status quo that seems more frozen in the 1950s before the advent of the civil rights and women's rights' movements in the South and nationally than equipped to fairly service and deliver progress to all members of the cooperatives equitably," noted Wade Rathke, a labor organizer reporting in the *Daily Yonder*. The project was titled "Democracy Lost and Discrimination Found: The Crisis in Rural Electric Cooperatives in the South."

The project examined all available records on all 313 cooperatives in the twelve states of the South. Of the 3,051 supposedly democratically elected board members, 2,754 were men, 90.3 percent, while 297 members were women, 9.7 percent. This disparity in numbers existed even though women in the South outnumbered men, 51.1 percent to 48.9 percent.

Examining those in co-op leadership or governing positions in the South, the project found—where information was available and verifiable—that of 1,946 executives and board members, 95.3 percent were white. Of 90 in leadership positions, 4.4 percent were black. Of the more than 2,000 governing positions for which the project could compile information, only 6 were Hispanic, 0.3 percent of the total. These figures were in contrast to the overall population of the twelve southern states at the time—69.23 percent white, 22.32 percent black, 10.19 percent Hispanic.

The project found that half of the states—Arkansas, Florida, Kentucky, Louisiana, Mississippi, and Tennessee—had three or fewer African American board members, with Louisiana and Kentucky having only one and Arkansas, Mississippi, and Tennessee having only two. Although Florida counted almost one-quarter (24.1 percent) of its population as Hispanic and Texas more than one-third (38.6 percent), only one Hispanic board member was serving in Florida and five in the entire state of Texas.[4]

What the project found in 2016, Shirley Jackson had lived. Born in 1939 on a farm outside Gilmer, she was the youngest of six children—she had five brothers—in a hardworking family that raised cotton, peanuts, sweet potatoes, corn, and vegetables. Her father worked on the railroad during the winter.

Her parents were among the few African Americans in Upshur County who were allowed to vote. "They didn't have to count the beans in a jar or guess how many bubbles there were in a bar of soap like many of their friends did," she recalled.

When Upshur Rural Electric Co-op was organizing, a Gilmer insurance man named Walter Stewart visited Jackson's father. Stewart, a board member, urged her father to pay the five-dollar membership fee, which he did, although he was never involved with the co-op afterward. "When the lights came on, I couldn't believe it," Jackson recalled. "It was like the Beverly Hillbillies going out to Hollywood. That was our way of life back then."[5]

Growing up on a farm was enough to persuade Jackson that life in the country wasn't for her. She had taken piano lessons from age six, so when she got to Texas College in Tyler, her ambition was to pursue a musical career in New York. She modified her plans slightly when she met a Texas College football and basketball star named Henry Jackson. He became a coach; she became a public-school music teacher—and the wife of one of the first African American board members of Upshur Rural EC.

Upshur Rural was no longer the co-op she remembered as a child. "Back then," she said, "when they needed us, we heard from them. Otherwise, we weren't involved."

"Taking a walk down memory lane," Fred McClure recalled something similar in San Augustine. One of the oldest towns in Texas, his hometown was still one of the most racially segregated when he graduated from high school in 1971, one year after the local black and white schools finally integrated. The three doctors in town still maintained separate waiting rooms for patients into the late 1970s.[6]

"It was the state of East Texas at the time," McClure recalled. There was no sense of mission to serve the underserved that had motivated co-op pioneers decades before. Although McClure went on the co-op youth tour to DC the summer before his senior year, black families as a rule simply didn't participate in co-op activities. "It's not that we were discouraged from participating, but neither were we encouraged," McClure recalled. "The co-op was where you got your electricity, nothing more."[7]

McClure would become the first African American president of the Texas A&M student body, a summa cum laude graduate, and recipient of the Brown Foundation–Earl Rudder Memorial Outstanding Student Award. After receiving his law degree from Baylor University, he worked in the Washington office of US senator John Tower (R-TX), and in the White House for Presidents Ronald Reagan and George H. W. Bush.

McClure, who became the chief executive officer of the George Bush Presidential Library Foundation and then director of his alma mater's Leadership Initiative, wasn't the last African American to be featured in *Texas Co-op Power*, although it took more than a decade for the next. In the August 1983 issue, the magazine ran a short feature about Lencola Sullivan, "a popular and personable weather analyst for KTVV-TV in Austin." Sullivan, the first African American to be crowned Miss Arkansas, had been caught in a citywide blackout a few weeks earlier, the result of a power surge the municipal electric system couldn't handle. *TCP* lauded the young woman for her aplomb under pressure while the station worked to get her back on the air. Every decade or so after the Sullivan feature, a black person made it into the magazine.[8]

Beauty queens were big in *TCP*, particularly in the late 1950s and early 1960s. Young, fresh-faced women hailing from small Texas towns frequently graced page 1 after being crowned Miss Young America, Miss FFA Sweetheart, Texas Dairy Princess, Texas Turkey Queen, Miss Wool of Texas,

Miss Mohair of Texas, Miss Rodeo, or, beginning in 1957, Miss Texas Rural Electrification. Usually college freshmen or sophomores, occasionally high school seniors, the women were not only attractive in bathing suits and evening gowns but also highly accomplished. *TCP* readers learned that they were honor students, baton twirlers, cheerleaders, active FFA members, basketball players, musicians, and church choir members.

Sydney Slack of Perryton, the first Miss Rural Electrification, was a drum major and an all-district basketball player. Gretchen Niebuhr, a sixteen-year-old Gonzales High School junior, was a runner-up in 1957, the year Slack won. Not only did Niebuhr make straight As, but her interests ranged from drama to coon hunting.

In addition to the accomplishments of the young women, *TCP* readers learned their height, their weight, their hair color, and, as an anonymous *TCP* reporter euphemistically put it one year, "other interesting statistics—34–22–34."

They were accomplished, to be sure, but the young beauty contestants featured in *Texas Co-op Power* were adornments (as well as welcome relief from the cavalcade of older white males who dominated every issue). In a publication that, until the mid-1950s, identified married women by their husband's name, never their own, the status of women was inadvertently revealed in the opening sentence of a September 1977 article: "Almost 700 Texas rural electric leaders and ladies gathered in Houston August 3–5 for the Association of Texas Electric Cooperatives' 37th annual meeting."

In the fall of 1970, a group of women organized the Texas chapter of the Electrical Women's Round Table, composed of women who worked for electric co-ops and other electricity-related industries. There was a reason the new group's program topic for its first meeting was "Women in a Man's World."[9]

There was little acknowledgment through the years—in *Texas Co-op Power* or elsewhere in co-op world—of women like Mabel Bryan Morriss, who almost single-handedly willed Bowie-Cass EC into existence. There was little mention of Susan Lorenz, who with her husband, J. P. Lorenz, not only fostered at least fifteen children on the Lorenz ranch east of Stockdale but also fostered an electric co-op. When rural electrification became a possibility in 1939, she rode horseback to talk to neighbors for miles around, working to persuade them to sign up for electricity from the Guadalupe Valley EC. She later became a director.[10]

Few beyond Upshur County and environs knew that Delta Scales took over as manager of Upshur Rural when the co-op's first manager, her husband, Samuel Scales, died in 1944. One of only two female co-op directors in the nation at the time, she led the co-op through the difficult war years. She remained director until 1950.[11]

Swisher EC was an exception. Faced with "more than its share of difficulties" in the early days, including an unusually sparse area, skeptical farmers, and the looming threat of World War II, the Tulia-based co-op turned to "the feminine members of the community," as *TCP* put it in September 1956.[12]

Among the women on whom the fledgling co-op relied was Ruby Woods, the home demonstration agent at Tulia. Along with county agent John Palmore, she talked tirelessly about the advantages of REA power to every farm family she encountered. Their task was daunting, since many of the farmers she and Palmore approached had no interest in electricity. A number of them feared that electricity would set their houses afire.

C. D. Taylor, Swisher's first president, recalled one man who called him out of a meeting to explain why he couldn't sign up. "You'd have to cut the stuff off every night. By the time you get way out to my place to turn it on again, half the morning would be gone." Woods and other Swisher women kept talking, kept persuading; Taylor and others kept organizing. And the REA kept saying no because of the county's low density. "I ran out of fingers trying to keep count of the number of times we were turned down," Taylor recalled. On September 20, 1939, the REA finally said yes. Three of the women who had worked so hard at organizing were among the co-op's first directors.[13]

Forty-seven years later, in 1986, J-A-C Electric Co-op in far North Texas chose longtime employee Sarah Sears as the very first female manager of a Texas electric cooperative. As *Texas Co-op Power* pointed out in 1994, she was still the only female general manager in the state, with the exception of interim manager Janice Johnson at Nueces Electric Cooperative.[14]

Sears had begun her career thirty-seven years earlier, as the billing clerk for J-A-C (named for the three counties it serves: Jack, Archer, and Clay). She was one of only two office employees at the time. She would go on to work as the system's billing clerk, bookkeeper, office manager, and assistant manager before being named manager. She already knew most of the co-op's members, *Texas Co-op Power* reported, because from 1950 to 1954, she and her husband, Vaughn "Red" Sears, owned the area's small telephone company. She ran the switchboard for the service connecting the towns of Scotland, Windthorst, and Bluegrove and nearby residents. She served as general manager for sixteen years.[15]

Hilco EC hired its first female general manager in 2007. Like Sears, Debra Cole had been with the co-op for years, working first as cashier and then assistant general manager before being named GM. She admitted to being a bit nervous, "not because I didn't know the business, but because I had worked with most of these people for many years as a co-worker, and there were only three other female managers in the state of Texas."[16]

By 2020, five co-ops were headed by women, including the largest, Pedernales EC. Although more than forty women around the state were serving as co-op directors in 2020, every board remained majority male. Most were all male. None had more than two female members.

Eight decades after its founding, Swisher EC served members in parts of six West Texas counties (Swisher, Castro, Randall, Armstrong, Hale, and Briscoe) over an area encompassing more than 1,800 square miles. Its board of directors was composed of seven men and no women (although several women were included among the co-op's "key personnel"). Swisher's leadership was not unusual. Several African American members were serving on boards, as of 2020. There was no black CEO and never had been.

When Curtis Wynn served as CEO of Roanoke Electric Cooperative in North Carolina, the co-op was one of the few in the country with a majority-black board, majority-black staff, and a black CEO. Based in the small town of Ahoskie, Roanoke EC served a rural area challenged by an aging population, little or no growth, and extreme economic need.

The co-op was successful despite those challenges, Wynn told *Dissent* magazine, because of its involvement in programs that extended beyond electricity. Focusing on economic development, Roanoke sponsored a Sustainable Forestry program and an African-American Land Retention program, assisting landowners with everything from financial support to timber management. He expected the co-op to benefit its members the way a co-op had benefited him.[17]

Wynn had learned about electric co-ops at an early age. Growing up in the Florida Panhandle, he had a job as a teenager cleaning large silos used for storing peanuts. When a job came open at West Florida Electric Cooperative in 1981, he jumped at the opportunity to emerge from those silos and learn to be a lineworker. After college at Troy State University, he returned to WFEC and in 1997 took over as president of Roanoke EC. He was the first African American to run an electric co-op anywhere in the country.

Members had to be at the center of the co-op's every activity, Wynn believed. With that objective in mind, Roanoke held monthly "Straight

Talk" forums in each of the counties where it operated, eliciting feedback from members and learning what was working and what wasn't. Growing out of the forums were a number of energy-efficiency programs, along with rural development funds from the US Department of Agriculture. Federal funding for energy efficiency allowed the co-op to provide upgrades to member-owners' homes and their own equipment at no up-front cost to either ratepayers or the co-op itself.[18]

Wynn also made sure his members—average annual income about $33,000—knew what was available in the here and now and what was on its way: energy-saving devices for the home and office right now, electric cars and trucks on their way to replacing fossil-fuel vehicles, not to mention fleets of electric buses hooked up to a huge electricity reservoir storing solar energy when the sun was shining and releasing it after dark.

Wynn told National Public Radio in 2021 that he was determined to make sure that the massive shift from fossil fuels to clean energy in the coming years would be equitable. It was fine for wealthy people to get their electric cars and energy-efficient technology, he said, as long as poor people didn't get ignored. Wynn's passion for getting every home in the Roanoke service territory hooked up to high-speed fiber-optic internet was in the rural electric co-op's New Deal tradition. "No one else would do it then," he told NPR. "And no one else seems to want to do it now. And here we are again."[19]

In 2019, Wynn became the first African American president of NRECA. He made diversity, equity, and inclusion at electric co-ops a priority, insisting that they had been cooperative values from the beginning. "This is not a new concept for electric co-ops," he told *RE Magazine*. "But as with any other part of our business, it requires co-ops to continuously assess and evaluate how we're doing, and to make adjustments and improvements as new circumstances arise."[20]

In Texas, Bill Hetherington, for one, was trying to put Wynn's concepts into practice. "It's not a numbers game," the CEO of Bandera Electric Cooperative told *RE Magazine*. "It's not, 'I have to go out and hire 6.8 out of 10 people with a Hispanic background because 68 percent of our community is Hispanic.' It's about treating employees with respect and equality. It's about promoting a culture of trust and integrity that allows employees to feel comfortable in voicing their opinions."[21] It was a work in progress at Bandera. In 2021, two women served on the nine-member board. At least one of the male members was Hispanic.

"I MAY GET SHOCKED INSTEAD OF LIGHT WHEN I GO HOME"

Growing up in the heart of the Midwest in the 1920s and 1930s, the youngster they called "Dutch" personified the all-American boy. Born in 1911 in the tiny town of Tampico, Illinois, he absorbed the small-town values of patriotism, hard work, and godliness. A lifeguard in the summer, a high school football player in the fall, and a dutiful son year-round, Dutch Reagan absorbed the small-town values of patriotic America.

The Reagans were poor and moved frequently. During the Depression years, while Dutch was away at college, his father supported the family by distributing government food and working for the Works Progress Administration, a New Deal effort to provide jobs for Americans desperate for work. Small towns where they lived organized electric co-ops.[1]

As a young man, Dutch was "an enthusiastic New Deal Democrat." He recalled years later that he was "a near-hopeless hemophiliac liberal," perhaps because he knew that FDR had helped rescue his family.[2] After college, he became a radio sports announcer, broadcasting football and baseball games throughout the Midwest, before drifting into show business. In 1937, Dutch was given a screen test by Warner Brothers and so impressed studio heads that he was signed on the spot. Ronald "Dutch" Reagan would go on to make more than fifty movies and countless radio and TV appearances. He also ventured into politics through his involvement with the Screen Actors Guild.

The committed liberal began drifting toward the center, although he remained a Democrat until concerns about Communists or Communist sympathizers among the Hollywood Left propelled him further to the right in the late 1940s. Although he voted for Dwight Eisenhower in 1952 and 1956 and Richard Nixon in 1960, he still considered himself a Democrat.

In 1954, his movie career seemingly at a standstill, Reagan signed on to host a weekly TV series called *General Electric Theater*. He also began making personal appearances around the country for GE, giving as many as fourteen speeches a day at conventions, Lions Club meetings, and business gatherings, all the while learning to apply his acting ability to his increasingly effective public-speaking skills. During his eight years with GE, his talks gradually evolved from congenial remarks about life in Hollywood into politics—assaults on Communism and Democratic Party domestic programs, seasoned with warnings about the evils of big government and big unions.

The Tennessee Valley Authority was one of his standard targets. He called it "a government power octopus"—until, that is, GE's chair reminded him that the TVA bought $50 million worth of GE equipment. Reagan dropped the octopus anecdote.

In 1962, GE dropped Reagan, perhaps because of concern that his steady march even further to the right might drive away customers. One biographer described his political evolution as "bouncing off the far-left wall all the way back across to the far-right wall."[3]

Although no longer GE's spokesperson, Reagan still traveled the country making speeches. The product he was pitching was himself, the handsome, congenial, compelling face of the Republican Party. (He had switched parties in 1962.) A year after GE severed ties with him, he happened to complain in a speech that, even though the task of getting electric power to America's farms had been accomplished years earlier, the REA "does not go into retirement." The co-op people let GE know they were not happy with that remark; GE let them know that Ronald Reagan was no longer its spokesperson. "The REAs are valued customers of the General Electric Co.," GE added.[4]

Reagan's glib remark should have been fair warning to electric co-op members when he became the GOP standard-bearer a couple of decades later. Nevertheless, rural Americans, including co-op members, were willing to ditch the incumbent Democrat, a Georgia peanut farmer who was well acquainted with electric co-ops.

In a speech to NRECA in 1977, President Jimmy Carter reminded co-op leaders of his REA connections. "I was thinking, during lunch, when I was contemplating coming over here to meet with you," he said, "that perhaps the most exciting and gratifying days of my life were when they turned on the electric lights in our house, when I was 13 or 14 years old. . . . Electric cooperatives have always been close to me, and to my family, as you well know. Those of you who know anything about my background realize this. The formation of the REA during the '30s opened up a new opportunity for an expanded and productive life."[5]

As Ted Case points out in *Power Plays: The U.S. Presidency, Electric Cooperatives, and the Transformation of Rural America*, "Ronald Reagan was the first president since the advent of the REA who had only an abstract notion of electric co-ops, a combination of his big-picture style and his motion-picture background. Hollywood was a long way from rural America, and California's electric co-ops were bit players in the Golden State's giant energy scene, a role they gladly embraced."[6]

Despite his connection with rural America, Carter was back on his Plains, Georgia, peanut farm three years later, denied a second term by the voters. His successor had the REA in his sights from the beginning. Allied with a Senate controlled by Republicans for the first time in twenty-six years, the new president was determined to make major changes to government loan programs, including the REA.

Granted, Republicans weren't the only REA skeptics. As far back as the Kennedy administration, some Democrats were questioning whether the REA had outlived its purpose. The skeptics included "influential Democrats," Norman M. Clapp, NRECA head during the Kennedy and Johnson administrations, recalled in a 1969 oral history interview. "Some of the old-timers still look upon the REA program as strictly a farm electrification program," Clapp said. "Once you hook up 98 percent of the farms according to that school of thought, you've done your job—nothing more to do; why are you still around now? That's not confined to either party. That viewpoint exists in both parties. Occasionally—well, I shouldn't say occasionally—among the career people within the [Johnson] Administration there are many who feel that this is a New Deal program that was fine in its day but has outlived its usefulness."[7]

"Anyone who visited small rural towns or farms at the beginning of the 1930s is not likely to forget the pervading odor of coal oil used in the lamps that were the only sources of light," the *New York Times* noted shortly after

Reagan was elected. A half century after the REA's founding, that odor had dissipated. In overwhelming numbers, rural Texans were switching their allegiance to the Republican Party just as Reagan had done. They may have been electric co-op members, but the co-op was simply where they got their electricity. The fact that Republican administrations had traditionally tried to hobble the REA—if not put it out of its misery—was of little concern. Unlike Sam Rayburn, Lyndon Johnson, Bob Poage, and lesser-known co-op pioneers around the state, they no longer considered rural electrification a mission.[8]

Reagan's anti-REA position was standard Republican fare before he was elected president. In 1957, President Dwight Eisenhower had pushed Congress to double the 2 percent interest rate on all REA loans and to cut in half the total amount the agency could lend in any year.

The man who had served as vice president during the Eisenhower administration launched an all-out attack on the REA once he became president. Shortly after Christmas 1972, a few months after President Richard Nixon's reelection, his administration directed the US Department of Agriculture to announce that as of January 1, 1973, the REA would no longer be receiving funds to be loaned at the 2 percent interest rate, even though the funds had already been authorized by Congress.

Nixon's plan was to ditch the federally funded REA loans and replace them with commercially backed loans at 5 percent interest. The president insisted his proposal would save more than $200 million in federal spending. What he didn't say was that the rural co-op enterprise would be gutted. The federal funds already allocated to electric and telephone co-ops would be impounded. In addition, the Nixon plan would more than double the interest rates future borrowers would have to pay. Among numerous loans left in the lurch by the surprise announcement was a $553,000 loan to a Texas co-op—perhaps, *Texas Co-op Power* noted, one of the last such loans ever made.

Robert D. Partridge, the NRECA general manager, knew immediately that the very existence of the co-ops was under threat. He warned that if the White House prevailed, more than a thousand electric co-ops would be forced out of business. Millions of consumers who depended on the co-ops would be hard hit, as well.[9]

Partridge urged co-ops to descend on Washington. On January 23, three weeks after the USDA's announcement, 1,400 electric co-op representatives from forty-six states (including 135 Texans) traveled to a Rural

Electric Rally. They were determined to persuade Congress to restore the REA direct loan program.[10]

Lawmakers responded. They passed legislation that would allow the REA to extend loans in the full amount authorized each year and also established a direct loan program. Both bills quickly passed both houses of Congress. On May 11, 1973, exactly nineteen weeks after first issuing the order that might have destroyed the REA, Nixon signed the new legislation. Ironically, the modified direct loan program called for in the legislation resulted in even greater co-op financing.

The REA would face additional threats in years to come, but for the moment its existence was secure. The same could not be said for the Nixon administration. Fifteen months after securing the REA's future, a scandal called Watergate would cost Richard Nixon his presidency.

A couple of months before Reagan was elected in 1980, NRECA head Partridge had received a letter from the GOP nominee. In the letter, the candidate assured the organization that, if elected, he would not change the rural electrification program without first consulting its leaders. That's not what happened, Partridge claimed.

Without consulting NRECA, Reagan proposed an even more drastic plan for reshaping the REA than anything his GOP predecessors had proposed. In an effort to get the federal government out of the loan business entirely, he wanted to prevent electric co-op access to the Federal Financing Bank and instead have co-ops go into the private money markets.[11]

Thirty-four-year-old David Stockman, the newly appointed director of the Office of Management and Budget, was Reagan's point man. A former congressman and supply-side true believer, the budget wunderkind was prepared to ferret out $40 billion in budget cuts from a list of between five hundred and a thousand federal agencies, a list he had memorized. He was prepared to "zero out" numerous programs and agencies. The REA was on his hit list.[12] "The whole countryside had been strung with electric wires by 1950, bringing the original mission of the REA co-ops to an end," Stockman contended. "But the co-ops had never quite gotten over the thrill of cheap electric power."[13]

Having grown up on a farm in western Michigan, where he was an active FFA member, young Stockman knew that co-ops had been relying on the US Treasury Department's Federal Financing Bank since 1937, thanks to an act that also established the Rural Electrification and Telephone Revolving

Fund. The act had given the REA authority to guarantee or insure loans from those sources.

The administration sought to end all 2 percent interest loans from the revolving funds and to reduce the total amount the revolving fund could lend in a year to $700 million in 1982, from $800 million. None of the actions would have reduced direct government spending, since all were "off-budget" functions that were designed to be self-financing.

In 1980, the REA guaranteed more than $6.6 billion in loans to the co-ops. Of this amount, $5.7 billion was for generation and transmission construction and went out through the Federal Finance Bank at market interest rates. The remaining $925 million was for expansion and improvement of existing systems and was provided through the Rural Electrification and Telephone Revolving Fund at 5 percent for most borrowers and 2 percent for small cooperatives. The revolving fund was replenished each year with principal and interest payments from borrowing cooperatives and was not included in the federal budget.

"The REA has substantially accomplished its purpose of providing concessional financing to make available electric and telephone service to rural areas," Stockman wrote as part of several proposed revisions of the 1982 budget.

Partridge characterized the Reagan plan as "good news for bankers and bad news for rural Americans." He warned that the changes meant that co-ops would have to pay higher interest on their loans and would find it harder to obtain the capital they needed to meet future needs. "The announced plan," he said, "holds the certainty of higher electric bills for 25 million people served by rural electric systems, because it is inflationary and will boost co-ops' operating costs. But it will not save the government money."[14]

Congressional Democrats, although outnumbered, tried to come to the REA's aid. They labeled White House attempts to reduce the available loan funds "a shameful effort."[15]

Barely one hundred days after his inauguration, Stockman's boss won a huge budget victory in the House. Sixty-three Democrats, including many with electric co-op constituents, joined their Republican counterparts in support. Congress passed his budget virtually intact.[16]

Amazingly, the REA survived Stockman's surgical incisions, but its battle with the Reagan White House continued. After the president's landslide reelection victory in 1984, including overwhelming support from grassroots co-op members, the administration set in with renewed resolve to dismantle the REA. It never happened.

Stockman left the Office of Management and Budget in 1985, frustrated that he had been unable to zero out the REA. "Four years and about 15 billion wasted REA dollars later," he wrote, "the fight would still be going on."[17]

In theory, Stockman was right about the fight continuing, but President George H. W. Bush, preoccupied with foreign affairs, never brought the same ardor to the battle as his predecessor. The REA got to catch its breath for four years. Ironically, it was the administration of the man from small-town Arkansas, presumably the REA's friend, who came close to killing it.

Bill Clinton did not cede the rural vote to his Republican opponent or to the formidable third-party candidate in the race, H. Ross Perot. The youthful former governor of Arkansas charged that the Reagan and Bush administrations "completely neglected the needs of rural Americans. The way they have run the REA is conclusive proof."[18]

Although Texas voters stuck with fellow-Texan Bush, rural voters nationwide gave Clinton 43 percent of the vote. "This was really a rural revolution," said NRECA chief Bob Bergland, a former secretary of agriculture.

Clinton friend Carl Whillock, president of Arkansas Rural Electric Cooperative, campaigned for the president-elect in Iowa, where he heard an "overwhelming" plea for jobs. "That's what caused the majority of them to vote the way they did," he told a Washington reporter for co-op publications.[19]

The man from Hope (Arkansas) immediately set out to prove he was no tax-and-spend liberal. On February 17, 1993, in his first State of the Union address, he told a joint session of Congress that he was focused on reducing the deficit. He announced his intention to cut federal spending by $246 billion by eliminating programs no longer needed. "I recommend that we reduce interest subsidies to the Rural Electrification Administration," the new president said. "That's a difficult thing for me to recommend. But I think that I cannot exempt the things that exist in my state or my experience, if I ask you to deal with things that are difficult for you to deal with."

Here we go again, co-op leaders around the country were thinking as they listened to Clinton's speech. Jim Morriss, TEC's general manager, wrote in the next issue of *Texas Co-op Power* that he understood the president's "share the pain" rationale, but he maintained that, basically, Clinton didn't know what he was talking about. "The phrases used to characterize rural electrification as no longer needed are the same tired, shallow statements that show more than anything that the originator of the phrases was uninformed," Morriss complained. "In short, we think the president has been given bad advice on cooperatives. The proposed cuts are projected to save $374 million over four years. While that amount is small relative to the total deficit, we are ready to do our part if fairness is used in deciding on cuts."[20]

Clinton heard similar complaints around the country. In remarks to the Cleveland City Club, he joked that after his proposal to cut the REA, "I may get shocked instead of light when I go home."[21]

The president, though, had found a winning issue. Debate about the REA's future lasted throughout 1993, with its leaders gradually realizing that times had changed since the agency's creation during the Depression. "The leadership of the '30s is all gone," Bergland admitted to the *Washington Post*. "It changes the politics of the REA, and I understand that."[22]

For only the fourth time in fifty-eight years, the statutory interest rate for insured loans for electric and telephone cooperatives was adjusted. In the beginning, loans were made at the same rate that money cost the government, below 2 percent. In 1945, the 2 percent rate was set by statute in return for an "area coverage covenant," which meant co-ops would extend service to anyone in their territory who wanted electricity. In 1973, the statutory rate was increased to 5 percent.

Nearly twenty years later, the House and Senate Agriculture Committees, working closely with NRECA, crafted a bill that reduced the cost of the REA lending program. The bill abolished the 2 percent loan program, replacing it with a 5 percent hardship loan available to co-ops serving the poorest and most remote areas. As part of a reorganization of the US Department of Agriculture, the REA became the RUS, Rural Utilities Service. "Most of us would have been more comfortable if no change had been required, but that was not an option," Bergland said. Glenn English, then an Oklahoma congressman chairing the Agriculture Committee, looked on the bright side. "We made lemonade from lemons," he said.[23]

Clinton won a second term in 1996, again with support from the rural vote. Electric co-ops managed to fend off deep cuts in their programs because of strong allies on Capitol Hill, but also because a bigger issue was looming. It was called "restructuring," and it was designed to allow large industrial users to shop for lower electricity prices. Most electric co-ops were wary.

"IT'S NO MORE BUSINESS AS USUAL"

"Regular readers of *Texas Co-op Power* know that we usually steer clear of politics and public policy issues on the feature pages of the magazine," publisher Peg Champion noted in the March 1999 issue, "unless it's an issue so vital to our readers that we just can't ignore it."[1]

The unignorable issue as a new century loomed was deregulation, a profound restructuring of the electric utility market that would allow consumers, businesses, and major industrial users to pick and choose their power source. Electric energy suppliers, including co-ops, would be able to compete for their business. One industry observer called deregulation "the biggest change in the North American electricity business since Thomas Edison began selling electricity in New York in 1882."[2]

As Champion pointed out, electric co-ops had been traditionally wary of deregulation. They remembered the old days when utility companies thumbed their city noses at potential rural customers. "Co-op members know they're still not the customer of choice; they're too far out and too spread out for energy providers to make money on them," Champion wrote. She also noted that residential and small commercial users still used much smaller amounts of electricity than large industrial customers. She explained that the restructuring schemes that lawmakers were considering would allow co-op members to choose whether they wanted their co-op to enter the energy marketplace and that whatever the individual co-ops decided to do, they would continue providing the same reliable service that had become their trademark.[3]

Mike Williams, CEO of Texas Electric Cooperatives, assured co-op members that the sponsors of a comprehensive restructuring bill, state senators David Sibley (R-Waco) and Ken Armbrister (D-Victoria), could be trusted.[4] Sibley, a former Waco mayor and the bill's primary sponsor, would be "a steady hand at the wheel this session as the electric co-ops—and the lawmakers themselves—negotiate the tricky turns, blind alleys and unexpected slick spots of electric utility restructuring," Williams said. His colleague Eric Craven, TEC's senior vice president for government relations and legal affairs, considered the Waco Republican the right man for the job. "He had the mind for it," he recalled.[5] Armbrister, in Williams's words "a longtime friend of electric co-ops," had carried legislation in 1995 that deregulated the wholesale market.[6]

In Texas, where rates were tied by law to the cost of fuel—mostly coal or natural gas—the push to deregulate the power industry had its origins in the early 1970s. When electricity rates began to soar because of rising natural gas prices and the OPEC oil embargo in 1973, anxious lawmakers decided that regulation of the vital industry was necessary. Under the scheme they came up with, electric utilities operated as regulated monopolies in specified geographic regions. Each utility was allowed to recover just and reasonable actual costs and a reasonable profit margin—city-owned utilities and some co-ops set their own rates—and each utility was obligated to provide all customers in the service area with reliable, safe, and reasonably priced electric service.

"We [the electric industry] are the last of the big monopolies that are being forced into open trade and open competition," Darryl Schriver, assistant director of government relations for TEC, said at the time. By the time Texas got around to the deregulation debate, eighteen states had deregulated and others were in the process. Lawmakers at the federal level also were considering various deregulatory schemes.[7]

Behind the push to deregulate—or restructure, as Sibley preferred to call it—were the big industrial users of electricity. They had been adamantly opposed for years, but then they came around to the position that a regulated monopoly was economically inefficient and that it skewed the relationship between product and price. Giant petrochemical plants on the Gulf Coast, lumber mills in the Piney Woods, and other voracious energy users wanted to be able to purchase electricity directly from a wholesale marketer. They would pay the local distribution utility to transport the power over its lines in the same way that long-distance phone carriers at the time used the local phone network to complete long-distance calls.

Texas, the largest producer and largest consumer of electricity in the nation, obviously was a key component in the radical restructuring. *Texas Co-op Power* reminded its readers that the Lone Star State had begun regulating the electric utility industry in 1975, when the legislature created the Public Utility Commission to set standards and rates for both electric and phone service. Before 1975, cities were primarily responsible for regulating electric rates.[8]

The big IOUs, suppliers of slightly more than 80 percent of all electricity in Texas, were eager to market their product anywhere and everywhere at competitive prices. Recognizing that they would have to win the hearts and minds of residential consumers before they got the undivided attention of the Texas legislature, they argued that consumers in an open marketplace could save between 15 and 30 percent on their monthly electric bills. A few months before the deregulation debate began, their lobbying group, the Association of Electric Companies of Texas, launched a $1.1 million statewide TV ad campaign touting the advantages that competition would bring to the residential consumer.

A Houston-based pipeline company named Enron worked to persuade Governor George W. Bush that deregulation was in the best interests of business, industry, and the consumer. Enron, a company that soared into the stratosphere and then came crashing, Icarus-like, back to earth in one of the most spectacular failures in Texas business history, already had a deregulation ally in the Governor's Mansion. Enron CEO Ken Lay had the governor's ear, as did Pat Wood III, a PUC commissioner who touted the efficiencies that competition would bring.

In 1995, Bush had recruited Wood to lead the PUC. The governor expected his appointee to deregulate the electricity market so that customers would be free from monopoly control. Airlines had deregulated, to the customer's benefit. Bush wanted something similar for electricity customers.

Years later, Wood recalled a conversation with the governor. "Y'all might think you have a lot of other jobs as a regulator," Wood told an Austin audience in 2022, recalling Bush's words. "But first and foremost, your job is to create a foundation for the state of rich economic development."[9]

Deregulation proposals had been bandied about for several years, but lawmakers got serious about the issue in 1995, when they instituted deregulation of the wholesale electric market. Legislative efforts to deregulate the retail market "were deader'n Elvis until last session," a utility lobbyist observed shortly before the 1999 session convened.

Efforts revived with a bill sponsored by state representative Mark Stiles (D-Beaumont) and state senator Jerry Patterson (R-Pasadena), both representing districts chock-full of large industrial users on the Gulf Coast. Their bill would have allowed full competition by 2002 and would have given consumers the ability to choose their electricity supplier. Despite the governor's support, the bill never made it out of committee. The state's electric cooperatives got credit—or blame—for quashing the 1997 effort. "We were the fly in the deregulation ointment,' Williams recalled.[10]

On a Saturday afternoon in late May, with the legislative session winding down, the governor summoned Williams to his office to talk about the co-ops' reluctance. The TEC president, with much more experience and knowledge about power issues than Bush, was unapologetic. "It was a very cordial meeting," Williams recalled years later. "He listened and tried to understand that the change was much too fast for rural Texas."

Craven, TEC's government relations director, also attended the meeting. He too remembered the governor being cordial. "That's just sort of the way he is," Craven said later.[11] He also remembered Bush's chief of staff, Joe Allbaugh, being somewhat less than cordial. "He sat over there not saying anything, arms crossed, a scowl on his face," he recalled, chuckling. (Allbaugh would go on to manage Bush's 2000 presidential campaign and then become his director of the Federal Emergency Management Agency, or FEMA.) Craven also recalled a Bush talking point during the meeting. "Kenny Boy really wants this," the governor said almost plaintively, referring to Ken Lay, the Enron CEO.[12]

Williams and Craven pointed out that the co-ops weren't the only ones with reservations. Many legislators couldn't get comfortable with such a major change. "The proposed bill created more questions than answers," Williams told *Texas Co-op Power*. "And there was not enough time to determine whether the major changes being proposed would benefit our co-op consumers."

The co-ops, with close to three million members residing in most of the state's 254 counties, didn't have the clout to single-handedly doom deregulation, but they had long memories that nurtured their skepticism. They owed their existence to the unwillingness of big utility companies to serve rural customers in the unregulated marketplace of the 1930s. As Robert Caro eloquently recounts in his LBJ biography, the lack of electric power essentially divided Depression-era America into two nations: "the city dwellers and the country folk."[13]

Even as the twentieth century gave way to the twenty-first, co-ops knew they were serving customers the big utility customers weren't terribly eager to entice. They knew that "customer choice" was likely to mean little for customers who were still too far out and too spread out. And they were concerned about the trend in other industries that had deregulated, where a few industry giants gobbled up the smaller companies, or existing companies merged, until only a handful of big companies remained in business. Customers were left with a choice in name only.

"Co-ops are not against competition, and we're not against restructuring," Williams told a legislative hearing in 1998. "But the fact is, deregulation for rural Texans has not been all that good, whether it's trucking or airlines or other industries that have deregulated. This is not about economic theory; this is about economic reality for rural Texans."[14]

"I am not personally an advocate of deregulation," Marcus Pridgeon, general manager of Guadalupe Valley Electric Co-op, wrote to his members shortly before the big legislative debate. "I do not believe it will benefit the 'regular Joe' consumer, and I do not believe it will benefit rural areas."[15]

Pridgeon's view had been the prevailing co-op view, and yet he and other co-op members were well aware after the 1997 legislative session that deregulation was not going to drift away like a lazy summer cloud over the Texas Gulf. After the session, House speaker Pete Laney (D-Hale Center) and Lieutenant Governor Bob Bullock directed committees to hold hearings and investigate the immensely complex issue. State representative Steve Wolens (D-Dallas), chair of the Senate Interim Committee on Electric Utility Restructuring, took the lead. He and Sibley convened public hearings around the state and visited other states that had already deregulated. Bush by then was focused on bigger plans for his own future.

Co-ops vowed to be prepared. Immediately after the session, TEC's board of directors put together an Industry Task Force to examine how various restructuring plans would affect the average co-op consumer. "Our objective was simple," Williams recalled. "If the legislature decides that restructuring is a good idea for Texas, how can we make it work best for our co-op consumers?"

Craven recalled a dawning realization on his part and by others that the co-op had leverage. If the IOUs were so desperate for deregulation, then the co-ops were in a position to shape the bill. "Why don't we write it?" he recalled suggesting to Williams. "Then it becomes a matter of selling it to the [co-op] membership."[16]

Craven talked to Sibley's staff, who told him their boss would never go for the idea. A few days later, a staff person called back. The senator liked the idea. "If you could get the co-ops and the munis onboard, you could work out the details later," Craven recalled. "If you have the template, then the others have to claw back. It's much easier that way."[17]

By February 1997, the restructuring task force had hammered out a list of basic criteria that would have to be met before electric co-ops could endorse any restructuring proposal. The co-ops wanted to see a business model in a restructuring bill that recognized the uniqueness of electric co-ops; they wanted co-ops to have opportunities to provide a range of services to their members; they wanted protections to avoid wasting resources; they wanted measures to preserve service reliability. Representatives from the state's eighty-four co-ops unanimously adopted the restructuring position at a February meeting in Austin.

Wolens, who had earned a reputation as one of the brightest House members as well as a tenacious consumer advocate, rekindled the legislative debate in December 1998 by prefiling an electric restructuring bill. His bill allowed residential and business consumers to choose their electric supplier as of January 1, 2002—if the PUC determined that adequate choice was available. A longtime friend of the co-ops, state representative Warren Chisum (D-Pampa) also filed a restructuring bill. The Wolens bill was something of a "shell bill," with details to be filled in later. Sibley's bill arrived with details filled in. If the Wolens bill was the opening tip-off, then Senate Bill 7 by Sibley, a former Baylor basketball player, was an early three-point shot.[18]

In an interview years later, Sibley revealed that his plan had its origins on a cocktail napkin. In 1998, he and Wolens, along with PUC head Pat Wood, were on a flight from California back to Texas after a fact-finding mission to see how the nation's other large, energy-rich state had crafted its deregulation scheme in 1996.

Jotting down notes and ideas on a napkin as they talked, the three men realized that the Lone Star State had several advantages over the Golden State. They came to realize, for example, that about a fifth of California's energy came from hydroelectricity imported from the Pacific Northwest. Texas, instead of relying on imports, was the nation's leading energy producer. The three also thought they saw an opportunity to create a more efficient system than the one California had set up, and they expected Texas would be able to sell cheaper electricity than California. "Sibley would have to find a more workable, less onerous, system for Texas," KUT-FM noted.

California's system also included a mechanism to cap its electric rates if the market got too volatile. It appeared to Sibley that big California IOUs like Pacific Gas & Electric could be vulnerable under certain market conditions. He knew their Texas counterparts would oppose such a system. As the recipient of thousands of dollars in campaign contributions from the IOUs, he was familiar with how they thought.

As the 1999 legislative session loomed, Sibley, Wolens, and Wood worked to achieve consensus among the major players. In his KUT interview, Sibley recalled getting them to commit to a bill sight unseen.[19] "At that point, they didn't care, really, what else it did as long as they weren't in it. So, they pledged and nobody knew," he said. "I mean, we did this, I remember it like during the Christmas holidays. Nobody was in the Capitol. So, they came to my office individually. I told them what the deal was."[20]

SB 7 was drafted in the basement of the Texas Senate, Sibley told KUT. Left out of the loop was the Texas Legislative Council. As the state's legal counsel for lawmakers, the council's job was to examine drafts and suggest changes. "I didn't trust them," Sibley said. "I felt like if we used the Texas Legislative Council, it would be leaked to the various lobbyists. And people would have a chance to start sniping at it before we ever even released it."[21]

In his State of the State address on the opening day of the session, Bush said he "look[ed] forward to working with Senator Sibley and Representative Wolens on an electric deregulation proposal that will cut costs for consumers while making sure electricity is available and reliable." The governor had no idea that Sibley had already drafted the plan. Nine days later, he unveiled his handiwork on the Senate floor. Lobbyists for the investor-owned utilities in attendance didn't know what hit them, the senator recalled.[22]

The process began in earnest on a January afternoon in 1999, when the capitol crowd gathered outside the Senate chamber. Gabbing and glad-handing, sharing stories and rumors, they were waiting for a press conference Sibley had called to unveil his bill. "There's more 'suits' here than you can shake a stick at," an environmental lobbyist mentioned to a *Co-op Power* reporter as the lobbyist surveyed the crowd. The "suits"—capitol jargon for dark-suited lobbyists—represented power marketers, municipal utilities, environmental organizations, consumer groups, chambers of commerce, energy companies, state universities, big energy users, electric co-ops, and numerous other interest groups. They knew that the comprehensive electric restructuring proposal the Waco Republican was sponsoring affected literally every Texan—every industry, every business, and every residential consumer of electricity.[23]

The massive doors opened, and the curious crowd shuffled into the ornate Senate chamber. As the tall Waco dentist began his remarks on the Senate floor, he was flanked by two supporters of the bill, one representing municipally owned utilities and one representing the co-ops—Mike Williams himself. More than a dozen lobbyists representing investor-owned utilities were lined up in the back of the Senate chamber. On the wall behind them were two huge paintings, iconic images that had hung there for years. One depicted the Battle of the Alamo, the other the Battle of San Jacinto. One portrayed a defeat, the other a victory. The IOU lobbyists weren't sure which would be emblematic in the deregulation battle ahead.

"He [Sibley] insisted that we not tell anyone we were working with him prior to that day," Williams recalled. "I remember a lot of very surprised looks in the audience, especially since we got credit or blame for killing dereg in 1997."[24]

"Where do the investor-owned utilities stand on this bill?" a reporter asked the senator.

"I don't know," he said. "I haven't talked to them. The people we talked to are the munis and the co-ops—and just about their sections of the bill. If you think we were trying to keep a secret, you're damn right we did. We were trying to craft a bill that would benefit the state of Texas. Now the process begins."[25]

The IOUs didn't just stand back and let it happen. They knew they had a lot to lose. At the time, Dallas-based Texas Utilities and Houston Lighting & Power controlled 55 percent of energy generation in the state. The Association of Electric Companies of Texas (AECT) represented the state's seven largest utilities. The companies had a combined worth of $70 billion in assets. They put those assets to work in 1999 (just as they had two years earlier) to try to kill the deregulation efforts.[26]

A few weeks after Sibley's press conference, Texas Utilities executive Tom Baker testified before a Senate committee on behalf of AECT and compared the large market share the IOUs controlled to organic monopolies. He maintained that sometimes dominance within any market simply evolves. "For an example, [in] the canned soup industry, the largest competitor has 73 percent of the market; lightbulbs, 66 percent of that market. We talked about electricity being a necessity, and I would just point out if we're talking about necessities, the largest competitor in the beer industry is 48 percent."[27]

Baker sought to persuade committee members that the state's plan would unfairly cap the market share of those utilities in a way that could cripple them. Even more important, he maintained, the cap from the state would be more stringent than the federal government's limits on market power.

Despite their opposition, the utilities at the bargaining table were aware that SB 7 was going to pass. They simply sought to protect their interests.

Eliminating the traditional monopolies, SB 7 split the industry into three divisions: generators, the division that produced the energy; transmission companies, which moved the electricity to consumers; and retailers, who bought electricity on the wholesale market and resold it to residents and businesses. Transmission and distribution remained regulated by the PUC. Generation would be deregulated and opened for competition. Power companies no longer would have total control of various geographic markets.

Sibley proposed freezing the electric rates IOUs charged until January 1, 2002. At that point, the rates would be cut 5 percent, and competition would begin. Competitors would be able to charge less than the utilities' rates, and, in Sibley's words, "consumers will be able to do much better than 5 percent." The freeze affected approximately 75 percent of the state's 8.6 million ratepayers whose energy was provided by IOUs. Consumers who chose not to shop around could stay with their own utility company and would get the 5 percent rate reduction even if they didn't switch providers.

Sibley's bill also tried to protect against one company playing a latter-day John D. Rockefeller and remonopolizing. Utilities with more than 20 percent of the generating capacity in a particular region would be required to file a plan with the PUC explaining how they intended to get below 20 percent. Either they could voluntarily divest or they could sell generation-production through a PUC-approved auction.

"This bill recognizes that municipally-owned utilities and electric co-ops are unique," Sibley said. "The rate cuts are not mandated for municipal and co-op customers unless their governing boards decide whether to open their markets to competition or they begin to compete outside their designated territories." The senator added, "Municipally owned utilities and electric cooperatives will be free to make choices that give their customers the best possible deal for electric service. One of our goals is to make sure that every customer benefits. We will not be a party to a bill where only a few benefit."[28]

The co-ops were pleased with Sibley's bill. They also were pleased that the veteran senator had made an effort to enlist their support before introducing it. "It was a masterful stroke of leadership," a lobbyist for Houston Lighting & Power told the *Austin American-Statesman*. Another told the *Houston Chronicle* that it "stings pretty good."[29]

According to Williams, Sibley's bill contained pretty much everything the co-ops needed. "We believe it's a restructuring plan that could ultimately benefit all consumers," he told *Texas Co-op Power*. Of course, Williams was well aware that much could happen before the end of the session, as state senator David Cain (D-Dallas) emphasized when he pointed out, "This bill is a work in progress."[30]

Reporting on the bill's progress in his regular *Texas Co-op Power* column, Williams reminded co-op members of the stark distinction between co-ops and IOUs. Co-ops, he noted, existed not to earn a profit for distant shareholders but to provide a service to their communities. Echoing the populist ire of early-day battles, he jabbed at the IOUs and their disdain for public service: "It's a concept foreign to the likes of Don Ingle, an executive with Cinergy Corporation, one of the largest electric utilities in the nation. 'We will focus our resources on those customers that will provide the targeted rate of return expected by shareholders,' Ingle remarked recently. 'In a deregulated market we shouldn't try to keep all customers. After deregulation is a reality we will . . . no longer have an obligation to serve.'"[31]

SB 7 made its way to the Senate floor on St. Patrick's Day 1999—"after seven hours, 51 amendments and years of work," KUT noted. Tired lawmakers listened while the bill's sponsor talked about beans. "Starting in 2002, people will be able to shop. If they don't like the electric provider they've got, they can switch," Sibley reminded them. "If the price of a can of beans goes up 10 cents, people shop somewhere else. If the price of electricity goes up, people for the first time will have a choice on what they're going to do. It's no more business as usual."[32]

Despite their opposition, the IOUs knew that SB 7 was going to pass, although they pressed Sibley to address what they considered crucial issues, including so-called stranded costs. These were inefficient or outdated investments—coal-fired or nuclear power plants, for example—that the utilities insisted had to be paid off. They amounted to about $4 billion, according to legislative estimates.

Most environmental groups supported the bill, albeit warily. The AARP and other advocates for older and low-income Texans also said yes after insisting that the state set up a plan to assist the poor with their bills. Ultimately, Texas Utilities and Houston Lighting & Power cut a deal, as well.

Still, Sibley and his cosponsors had concerns as the St. Patrick's Day vote loomed. State senator Mario Gallegos (D-Houston) asked Sibley whether, under SB 7, his constituents "would be receiving the same level of services . . . when [a] hurricane . . . or a storm comes in."

"Senator, I think you can represent that to your constituents without doubt," Sibley said.[33]

State senator Drew Nixon (D-Carthage) implored senators to consider the possibility of unintended consequences. "Because we're talking about changing something that every citizen of the state needs to have good, reliable, reasonable-priced access to," he said. "And I question whether there's such a big problem with our current system that we need to make that kind of change."[34]

Nixon's colleagues did not share his concerns. The full Senate easily passed Sibley's bill on a routine voice vote. The House devoted one more committee hearing to SB 7 and then amended it sixty-nine times before final passage. Governor Bush signed the bill in June 1999.

"Texas now joins 24 states that have restructured their electric industries, either by regulatory orders or through legislation," Williams wrote in his *Texas Co-op Power* column. "But unlike some other states, Texas didn't go off the deep end. The Texas bill is a middle-of-the road approach to restructuring. It borrows some of the best ideas from other states and transforms them into a plan for Texas that places a strong emphasis on flexibility and local control, offering the best chance for co-op customers to benefit from a restructured industry."[35]

The change, Williams explained, meant that co-ops went from managing merely an electric utility to managing risk. He offered an example: "Ten years ago, the electric cooperatives in the Texas Panhandle joined together to form a generation and transmission cooperative, Golden Spread Electric Cooperative. The new cooperative was formed to help manage power contracts and, because of its size, it received a better rate than an individual cooperative ever could."[36]

Golden Spread's general manager, Bob Bryant, had put together a coalition that not only built a state-of-the-art power plant, Mustang Station, but also worked out a way to keep the power flowing when there was a greater-than-normal need for electricity. "In other words, the customers of the electric cooperatives served by Golden Spread don't have to worry when an ice storm or a heat wave buffets their region," Williams said. "The co-op is prepared."[37]

Years later, Winter Storm Uri would severely test that optimistic assessment. Golden Spread's management and members worried, but the G&T survived. It was, indeed, prepared.

The transition to a deregulated market didn't happen overnight, even after officially going into effect in 2002. The IOUs persuaded the PUC to back their plan for stranded costs as they dismantled their monopolies in the run-up to '02. Environmental groups were pleased to see the percentage of renewable energy rise in the state's power grid, and Texas became the nation's leading generator of wind power. The state's fund to help low-income Texans, known as System Benefit Fund, got going, before lawmakers dismantled it in 2016.

Phasing in deregulation to allow new competitors to enter the market, the state enforced a price floor on the former monopolies, barring them from charging rock-bottom rates to shut start-ups out of the industry. Those restrictions faded in 2007, once newcomers had time to get established and the energy market became deregulated.

A complex marketplace emerged, with private power generators offering their product for sale to retailers across the state. Only the transmitters remained regulated, since cities would not tolerate multiple private companies stringing power lines wherever they wanted.

Proponents promised that deregulation would make power cheaper for Texans. If customers could choose between rival electric providers, then those providers would become eager to do everything possible to reduce their costs and win more customers. That promise was never really fulfilled. After an initial burst of enthusiasm about the benefits of competition and consumer choice, most consumers around the state merely settled for the provider they had before deregulation.

"NO ONE WAS MINDING THE STORE EXCEPT THE GENERAL MANAGER, AND NO ONE WAS MINDING THE GENERAL MANAGER"

When Lyndon B. Johnson left Washington in January 1969, he was, in the words of Leo Janos, writing in the *Atlantic*, "a worn old man at 60, consumed by the bitter, often violent, five years of his presidency." It took him nearly a full year "to shed the fatigue in his bones."

One of the reasons Johnson decided against running for reelection in 1968, in addition to the albatross of Vietnam, was his awareness that heart ailments ran in his family. His father had died at sixty-two, and he doubted he would survive another four tumultuous years in the White House. "I'm going to enjoy the time I got left," he told friends.

Apparently, he did—as much, that is, as a complicated, mercurial man of deep feelings, opinions, and resentments could enjoy himself. Once he finished the obligatory memoir of his White House years, *The Vantage Point*, he threw himself into managing the 330-acre ranch on the Pedernales near Stonewall that he and Lady Bird had purchased in 1951. He happily got his hands dirty, installing a complex irrigation system and, wearing shorts, helping lay pipe in the river. He constructed a large henhouse, planted acres of experimental grasses, and built up his cattle herds by making shrewd purchases at the weekly Stonewall cattle auction.

Old friends came for dinner and the requisite tour of the ranch, LBJ behind the wheel of his white Lincoln Continental as it dipped and plunged like a pleasure boat over grassy pastures where fat cattle grazed. He took up golf, playing the municipal course at nearby Fredericksburg or on trips to Mexico. Each February, he took over a seaside villa in Acapulco and flew in friends, family, and aides. He let his hair grow long, to distinguish himself, an LBJ associate told Janos, from President Richard Nixon's close-cropped aides.

Time was running short, as LBJ well knew each time he threw back nitroglycerin tablets to quell continual chest pains. He took up smoking again—"I'm an old man, so what's the difference?"—and had to resort to an oxygen tank several times daily. Renowned heart surgeon Michael DeBakey examined him and decided that his condition was so tenuous he might not survive surgery.

On January 22, 1973, a cold morning in the Hill Country, Lady Bird Johnson noticed that her husband was preternaturally quiet but nothing seemed wrong, so she decided to drive into Austin to do some shopping. At 4:05 that afternoon, she got the call from the Secret Service via car telephone. Her husband had suffered a heart attack an hour or so earlier and could not be revived. LBJ was sixty-four, two years older than his father when a heart attack claimed him.[1]

One day during Lyndon Johnson's presidency, NRECA administrator Norman Clapp was riding through the Hill Country with former Blanco County judge A. W. Moursund, LBJ's lifelong friend and business partner. Moursund was behind the wheel. Sudden static from the car radio startled Clapp, as did the voice of the president of the United States filling the car.

"The Judge has a radio in his car, which is hooked up to the President's communication system," Clapp recalled. "As we were driving along, the President's voice came on the intercom, and he was talking, I take it, to a foreman on the ranch about a fence that they were going to build. It was really quite interesting to hear the President of the United States with all of his burdens who was back on the ranch that day checking over the plans for this kind of work. He was telling the foreman to go ahead with what he proposed to do, but just be damned sure that the price was right on something or other that went into the construction. I have the feeling that he follows things in the federal government just about that closely."[2]

The man who had written letters of welcome to each new Pedernales member when he was a young Hill Country congressman wasn't as involved with the local co-op in his postpresidency years, and yet Johnson no doubt kept abreast of co-op business. Moursund was for many years

Pedernales EC's general counsel. "The Judge," as he was known throughout the Hill Country, exercised inordinate influence over all phases of the co-op operation.

According to LBJ biographer Robert Caro, Moursund was a collector of campaign cash for Johnson, surreptitious and otherwise. He was also a regular domino partner when Johnson spent extended time at the ranch. Caro described one game—with LBJ neighbors Wesley West, Gene Chambers, and Moursund—that "began in the morning, resumed after lunch, and then, after dinner at the West Ranch, went on there for several more hours."[3]

During his years in the White House, LBJ occasionally relied on Moursund to gauge the opinions of the man on the street. According to one LBJ biographer, Johnson called his old friend to see whether he could find Texans who opposed his 1965 decision to send troops to the Dominican Republic to forestall the establishment of a "communist dictatorship." Moursund told the president he had a PEC meeting in Johnson City to attend but would see what he could find out afterward. A few hours later he called the president back to report that he had found not one critic. The president was pleased.[4]

In the years following LBJ's death, the electric cooperative that owed its existence to Johnson's energy, ingenuity, and persistence was prospering, but it had begun to lose touch with its co-op roots. Pedernales Electric Cooperative "ran off the rails," a board member recalled years later. Four decades after LBJ's death, a reform slate of PEC board members and a new general manager would take control. They hired Navigant Consulting, Inc. to examine in detail the previous ten years of the PEC operation. A 390-page investigative report prepared by the firm in 2008 showed in damning detail just how far the pioneering co-op had strayed from traditional co-op practices and principles.

The co-op's business practices for nearly four decades would suggest that its executives and board members had never read, much less followed, the Seven Cooperative Principles regarding democracy, equality, equity, and solidarity, not to mention honesty, openness, social responsibility, and caring for others. It's not hard to imagine how Johnson, a man of volcanic temperament, would have reacted had he lived long enough to see what had happened to the co-op he had helped create for the express purpose of helping his Hill Country neighbors.

Like barnacles on a neglected vessel, questionable, indeed illegal, practices began to adhere to Pedernales in the mid-1970s. Disgruntled members and three small newspapers in the Pedernales service area—the *Wimberley View*, the *Marble Falls Highlander*, and the *Hays County Citizen*—regularly called attention to pervasive wrongdoing, but it wasn't until the first decade of the new century that members, the media, and two state legislators began to get results. Through the years, concerns centered on three men: the aforementioned Moursund, longtime PEC general manager Bennie R. Fuelberg, and longtime board president W. W. "Bud" Burnett. The three exercised near-dictatorial control over the largest—and arguably the most successful—electric co-op in the country.

Moursund was the owner of various businesses during his lifetime, including the Moursund Law Firm, Llano-based Arrowhead Bank, Cattleman's National Bank in Round Mountain, the Moursund Insurance Agency, and other land- and ranching-related businesses. A member of the Texas House of Representatives from 1953 to 1959, he was the co-op's general counsel as a full-time employee for more than thirty years (1951 to 1984) and as an outside provider on retainer from 1984 until his death in 2002. His son Will inherited the role of co-op general counsel through the family law firm, Moursund, Moursund & Moursund.

The elder Moursund founded Cattleman's National Bank in 1986. Lightly capitalized, it was controlled by the Moursund family after the death of the patriarch. The bank held nearly half of Pedernales's revenues, totaling $238 million in 2007. Two Pedernales board members served on the bank's board of directors. Fuelberg was an "advisory director."

The bank had been awarded the co-op's business without competitive bidding. The Moursund family's various enterprises also enjoyed no-bid arrangements to provide legal, insurance, and real estate services to the co-op. The investigation conducted by Navigant Consulting concluded that Moursund-related individuals and entities had received in excess of $4 million from their affiliation with the co-op over the previous twenty years. That amount included about $200,000 a year that the co-op paid the Moursund law firm. "Both current and former members of the Board, as well as PEC managers admitted knowing through personal knowledge that Cattleman's bank was owned by the Moursund family; that certain PEC Directors sat on the Board of Cattleman's bank; and that Mr. Fuelberg had once sat on the Board of Cattleman's," Navigant Consulting noted in its report. "However, there was no systematic briefing of, or review and approval by the Board of potential conflicts."[5]

W. W. "Bud" Burnett may have wielded even more power and influence than A. W. Moursund. "No one ever won an argument against Judge Burnett," his 2020 obituary in the *San Marcos Daily Record* recalled.[6]

A four-term Hays County judge, Burnett joined the Pedernales board in 1968 and served as board president for more than thirty years. In 1987, he also began receiving a paycheck as a full-time PEC employee with the title of "coordinator." With salary and benefits totaling $190,000 in 2006, Burnett was to "interface with legislative and regulatory bodies" and report directly to the board. Just how much interfacing the longtime coordinator did was difficult to determine, since Burnett had no office, no staff, no regular hours, and no specific duties.

"While Mr. Burnett did, in fact, appear to have some limited role within the Cooperative," Navigant Consulting noted, "his recruitment as Cooperative Coordinator in 1987 appears to have stemmed from his personal relationship with Mr. Fuelberg and his personal financial situation at the time, rather than as a result of the Cooperative's need for the Coordinator position as described."[7]

Burnett served in his dual role as coordinator and board president until he resigned from the coordinator position on November 30, 2007. He resigned as board president on January 18, 2008. "Although in his position as Coordinator Mr. Burnett reported directly to the Board, as a salaried employee of the Cooperative, Mr. Burnett was subject to the influence of the Cooperative's former General Manager, Mr. Fuelberg, who was in a position to exercise influence over Mr. Burnett both in his role as Coordinator and as Board President," the Navigant report concluded. "Mr. Burnett's status as an employee of PEC, under Mr. Fuelberg's supervision, created potential to impair Mr. Burnett's independent judgment in his role as Director."[8]

Everyone who had anything to do with PEC was "subject to the influence" of Bennie Fuelberg. Essentially the co-op's chief operating officer, Fuelberg had signed on with the co-op in 1972, the organization's tenth employee when he was hired. When he was named general manager four years later, the co-op's monthly expenses totaled $33,000. When he resigned in 2008, he was presiding over an organization of more than eight hundred on the payroll, with monthly expenses of more than $33 million.

Fuelberg was an autocrat, PEC employees and co-op associates told Navigant. He ruled with an "iron fist." Dare to question Bennie, employees knew, and you would either be without a job or, if you were lucky, you would merely be demoted or transferred to another position. He used internal transfers or the threat thereof to keep his department managers off balance and beholden to him. "Mr. Fuelberg was generally described as being unreceptive to the ideas of others, especially when they differed from his own opinions," the Navigant study reported.[9]

Under Fuelberg's oversight, budget and cost-control measures were essentially nonexistent, expense controls were lax, and employee complaints regarding abusive management practices were ignored. Fuelberg controlled the board, rather than the other way around. Its seventeen members functioned as his personal politburo.

Responding to Fuelberg's dictates, the board made bylaw revisions via telephone conference calls after discussing them at unannounced board meetings. No agenda was prepared beforehand, and no minutes were recorded afterward. Co-op member-owners were not allowed to attend.

Eric Craven, TEC's senior vice president for government relations and legal affairs, suggested years later that there might have been a legitimate rationale for closed meetings at PEC and elsewhere. "I think lawyers told them they were too open, that they ran the risk of being taken over," he said. "It just went too far."[10]

TEC general manager Jim Morriss noted in his *Texas Co-op Power* column for October 1993 that annual elections for co-op directors around the state "ranged from hotly contested (usually for some specific reason) through mildly competitive or uncontested to incumbents continuing to serve without being reelected because there was no quorum present at the meeting." He told the story of one co-op election that had to be repeated on the spot because of an unusual tie vote. Before the revote, four members left the meeting and two changed their votes. That meant the incumbent lost by six. "He couldn't help but observe," Morriss recounted, "that he would have won by two votes on the first ballot if his daughters had attended as he had reminded them."[11]

Pedernales during the Fuelberg era never had to worry about such uncertainty. The board election process was basically controlled by Fuelberg, Burnett, and Moursund, which meant that the board itself was self-perpetuating, with the same members serving three-year terms for decades. The board regularly met behind closed doors.

As the *Austin American-Statesman* explained in 2008, board members up for reelection were allowed to select the committee that nominated candidates. The committee met and posted the nominees shortly before the pro forma board-member election, held at the co-op's annual meeting. Results were a fait accompli, one reason why the average tenure of board members serving in 2008 was approximately seventeen years. (Burnett, with more than forty years, served the longest.)

In 2007, O. C. Harmon, who was up for reelection in June, appointed two men to the nominating committee, a longtime friend and his son's brother-in-law. The year before, the nominating committee was composed of the son of board vice president E. B. Price and the brother of voting director R. B. Phelps. Outsiders stayed that way: outside.

"They [PEC] have basically written the rules in such a way that it's really hard to allow any [board election] competition. The democratic process has been circumvented,'" Becky Morris, a former wind industry and PEC employee, told Claudia Grisales of the *American-Statesman*. "The members are supposed to own the co-op, not the few at the top."[12]

Board members either ignored or failed to notice Fuelberg and other top executives, with their wives, flying first class to San Francisco, San Diego, New York City, Las Vegas, Albuquerque, and other cities around the country. Their hostelries of choice included the sumptuous Jefferson Hotel in downtown Washington, DC, and either the Ritz-Carlton overlooking Central Park in Manhattan (rates at the time $595 nightly) or the Four Seasons, where their bill for 2005 amounted to $4,670. They also ate well. An outing to Morton's The Steakhouse, billed to the co-op, included seven ribeye steaks, twenty mini crab cakes, twenty salmon pinwheels, and three shrimp Alexanders costing fifty-nine dollars each. ("That's not Hill Country Cupboard grub," a former TEC employee mentioned, referring to a modest Johnson City eatery near PEC headquarters that claimed to serve "the world's best chicken-fried steaks.")

Pedernales members were doubtless unaware, but one year they treated Fuelberg and friends to a Celine Dion concert at Caesars Palace in Las Vegas, with tickets costing $2,000. Fuelberg also spent co-op money every year on expensive Godiva chocolates for distribution to select staff and favored visitors.

Investigative reporter Grisales of the *Austin American-Statesman* wrote that from 2002 to 2006, Fuelberg and his wife, one other executive and his wife—and in at least one case, a girlfriend—amassed approximately $700,000 in credit-card charges paid by the co-op.[13]

Testifying before a congressional committee in 2008, state senator Troy Fraser (R-Horseshoe Bay) noted that all board members were given free lifetime health insurance for themselves and their dependents, as well as free physicals worth $3,000 for members and spouses at the famed Cooper Clinic Health Spa in Dallas. The board also created policies allowing retiring board members to receive emeritus status with compensation of $1,500 per month for life, in addition to the free lifetime health insurance for both the member and dependents.[14]

"It appears the former General Manager effectively isolated the Board and everyone else in the Cooperative from having any meaningful oversight," Navigant observed. "No one was minding the store except the general manager, and no one was minding the general manager."[15] No one was minding the GM in large measure because Pedernales was thriving. In 2008, the co-op was the power company for more than 226,000 members in a twenty-four-county, 8,100-square-mile service area. The Pedernales service area included part or all of more than forty municipalities, with more than fifteen thousand miles of energized line and three hundred miles of transmission line. Annual revenues were more than $450 million; total assets exceeded $1.1 billion.

The Navigant study acknowledged that Fuelberg's passion for customer satisfaction was in part responsible for the co-op's success. Annual electric utility surveys conducted by J. D. Power & Associates consistently found the organization at or near the top in customer service and customer satisfaction, power quality and reliability, and communications. Navigant noted, however, that Fuelberg's single-minded focus on customer satisfaction, although admirable, also appeared to be at the expense of sound fiscal practice, including nonexistent budgets, lax expense management, and inattention to cost controls.[16]

Needless to say, Fuelberg made sure he was well compensated over the years. Appearing before a legislative committee investigating Pedernales in 2008, TEC general manager Mike Williams presented a survey showing that PEC's pay scale at the upper echelons was grossly higher than that of other member-owned co-ops in Texas. Williams noted that the average pay in 2006 for all co-op general managers in the state was $150,671. For the fourteen largest co-ops, excluding Pedernales, the average pay, benefit, and expense packages for general managers in 2006 totaled $224,380. Pedernales paid Fuelberg $871,000 that year. He also was the recipient of a five-year, $2 million deferred compensation package.

Pedernales board members, according to Williams's survey, also did well. The average pay for board members around the state, many of whom served part time, was $13,274. At Pedernales, board members each received $50,322 in 2006. Board members were also eligible for an "emeritus" program that offered lifetime pay and benefits. Below the upper echelon, Pedernales pay and benefits were in line with those of other electric co-ops.[17]

"No other major electric co-op in the U.S. comes close to the overall [board compensation and benefits] figure," the *Austin American-Statesman* editorialized in 2008, "which is about 10 times the board's pay at the second-largest co-op in the country, Jefferson, Ga.,-based Jackson Electric Membership Corp."[18]

Pedernales may have prospered during Fuelberg's reign, but his recommendations for investment in several ventures, most of them outside the co-op's core business, turned out to be extremely costly. They included a power generation venture in conjunction with neighboring Bluebonnet Electric Cooperative that was discontinued in the 1980s; an ill-advised venture to buy a New Mexico software company; a costly acquisition in 2000, with little due diligence, of Kimble Electric Cooperative; and an effort, ultimately abandoned, to develop Texas Skies, a broadband venture.[19] "While some of these ventures, if successful, might have provided benefits to the co-op's members, the costs incurred have outweighed any perceived or real benefits to the members," Navigant said.[20]

"A self-perpetuating board manipulated by an autocratic general manager who didn't consider himself accountable to anyone was a scandal waiting to happen, and it eventually did," the *Austin American-Statesman* editorialized in 2011.[21]

The rumbling volcano of rumors, worries, and dissatisfaction with Fuelberg and associates finally erupted on May 16, 2007, when Pedernales member Lee Beck Lawrence filed a $164 million lawsuit in Travis County District Court seeking class-action status against PEC and PEC's board and management. Lawrence alleged breach of contractual and fiduciary duties; management negligence; and excessive compensation, benefits, and expenses. He also raised questions about PEC's investment in Envision and the co-op's unwillingness over the years to offer its members capital credits (annual payments based on excess revenues over expenses). PEC was the only electric co-op in the state not to return capital credits to its members.

One co-op member called the lawsuit "a member uprising." In his congressional testimony, state senator Fraser noted the paradox that the members were basically suing themselves over perceived wrongdoing by the co-op and its board.[22]

Things began to change almost as soon as the investigative spotlight was switched on. The judge hearing the class-action case ordered that new elections be held under rules that opened up the process. Fifty-eight candidates filed for five open seats, and on April 30, 2008, Pedernales held its first-ever candidate forum. Members cast ballots online, by mail, and at the annual meeting. On June 21, 2008, five virtual outsiders won; the one incumbent in the race lost. Nine of the seventeen incumbent board members resigned. "We have, for the first time in many years, directors who have been elected in a fair and competitive election by members," new board member Patrick Cox told the *American-Statesman*. "We need to be a model on how cooperatives can operate honestly and fairly in the best interest of the members."[23]

Fuelberg and Burnett resigned in 2008. The new board selected Juan Garza, a well-regarded former general manager of the city of Austin's municipally owned electric utility, as the co-op's new general manager. (Garza had grown up in Cotulla, the small South Texas town where Lyndon Johnson had taught school before embarking on his political career.) Key management positions were filled by new people, as well.

In February 2009, PEC canceled a bonus retirement package for Burnett that the board had approved, at Fuelberg's behest, in 2001. That agreement would have paid Burnett nearly $10,000 a month in retirement pay, in addition to other benefits. The board halved the amount. "Until we got in there, we had no idea how bad this really was," recalled Cox, who was elected to serve out Burnett's unexpired term and who would eventually serve as board president. "That's one of the reasons we had to get Navigant involved. Just this litany of problems. We had this idea we were going in and get this solved within a year. No! It was impossible. It took years."[24]

Fuelberg and Burnett refused to cooperate with the Navigant investigation. Fuelberg destroyed computer files, and holdover board members tried to block plans by two Pedernales-area legislators—Fraser and state representative Mark Rose (D-Johnson City)—to conduct a state audit of the co-op.

Garza and the new board—reduced in size from seventeen to seven—immediately began to implement comprehensive reforms designed to reduce costs and improve member services, while transforming PEC into one of the most transparent and responsible electric co-ops in the nation. Board meetings were available for viewing on streaming video. Garza even provided the *American-Statesman*'s Grisales with her own desk and computer inside PEC's headquarters. Grisales had doggedly written dozens of stories about PEC's questionable dealings.

Reforms included a landmark Bill of Rights for member-owners, a document that underscored the co-op's commitment to open meetings and records, as well as fair, open, and democratic elections. A whistle-blower policy was implemented. Board members would subscribe to a code of ethics and agree to conflict-of-interest covenants. Their compensation was reduced substantially. The position of coordinator, Burnett's sinecure for thirty-nine years, was eliminated. Directors up for reelection no longer would help select the committee that nominated candidates. Candidates henceforth would be required to announce their candidacy at least seventy days before the co-op's annual meeting.

In years to come, PEC would commit to satisfying 30 percent of its electric generation capacity requirements from renewable generation resources, including distributed renewable generation. "It was well-received," Cox recalled. "Ranches in the service area already were using solar panels." The co-op met its goal within a few years. PEC adopted revised energy-efficiency policies, committing to a goal of annual reductions of 20 percent of future demand growth through efficiency and demand-management programs. The co-op met that goal, as well.

Meanwhile, the new regime continued to find questionable practices hidden over the decades by its predecessors. Shortly after Garza was named general manager, he informed the board that Texas Rangers had visited co-op headquarters in Johnson City to investigate a $565,000 bank account held in the name of the failed Texland venture and long considered defunct. Initially, the account was with Johnson City Bank but was transferred to Cattleman's National Bank when it opened for business in the 1980s; only Fuelberg and Burnett had access to it. The new board parceled out the money to the co-op's 218,000 members. The co-op also paid out capital credits for the first time in its history, in the amount of $23 million over five years. "PEC is now recognized as a model for all electric cooperatives," Garza declared at the 2010 annual meeting. Although Garza and the co-op would part ways soon thereafter, the reform initiative continued.[25]

Tom "Smitty" Smith was pleased. The longtime head of the consumer advocacy group Public Citizen, a crusty, bearded fellow known for his outspoken opinions, acknowledged in 2008 that the co-op had turned a corner. "Pedernales has stepped to the forefront and has one of the most progressive energy policies in the country," he said.[26]

In 2016, the US Department of Agriculture's Rural Utilities Service (RUS) awarded $68 million to PEC. The loan allowed the co-op to take advantage of the Energy Efficiency and Conservation Loan Program to make low-interest financing available to consumers for investments in efficiency and renewable energy. PEC used the RUS loans to create "Empower Loans," a program that offered members up to $20,000 for grid-tied distributed energy resource (DER) systems, including distributed solar photovoltaic systems and grid-tied battery storage systems. The loans could also be used for grid modernization to drive energy efficiency initiatives.

"The historic public-private partnership between the RUS and electric co-ops that led to an electric grid spanning the continent has evolved enabling rural communities to leverage the benefits of advanced technology," said PEC CEO John Hewa. "Even as the technology for delivering safe, affordable and reliable electricity is revolutionized with automation and advanced communications technologies, the end-goal remains the same: strengthening economic prosperity in rural America." Jim Matheson, CEO of the National Rural Electric Cooperative Association, added, "The PEC loan program illustrates how the cooperatives' consumer-centric business model can promote innovation and strengthen communities."

On March 10, 2008, the parties in the class-action lawsuit reached an agreement that provided for mutual releases of the parties involved in the litigation; the retirement of $23 million in patronage capital by PEC through bill-credits to then-current members; the payment of up to $4 million in attorney's fees and out-of-pocket costs by PEC in connection with the suit; agreed mutual support for the terms of the settlement; and the performance of an "independent internal investigation" by Navigant Consulting covering the ten years prior to December 31, 2007. The presiding judge approved the case on May 5, 2008, although several parties later appealed.[27]

In January 2008, Hill Country district attorney Sam Oatman launched a criminal inquiry into PEC. Eight months later, Texas attorney general Greg Abbott took over the investigation. In 2010, a Gillespie County jury convicted Fuelberg of third-degree theft, money laundering, and fiduciary misapplication of property during his more than three decades at the co-op. Authorities charged him with conspiring with Walter Demond, a lawyer for Pedernales, to funnel more than $600,000 in utility money through Demond's law firm to Fuelberg's brother, lobbyist Curtis Fuelberg, and about $58,000 to the son of a former Pedernales board member. The former general manager was sentenced to three hundred days in jail and five years' probation, while Demond was sentenced to five hundred days in jail. Fuelberg began serving his time behind bars in 2015, five years after his conviction. "Pedernales Electric Cooperative expects justice for its nearly 230,000 members," the co-op announced in a written statement at the time the disgraced general manager went to jail. "PEC is expecting the complete fulfillment of the consequences of Fuelberg's criminal actions against PEC and its membership. With Fuelberg having finally exhausted all appeals, PEC supports the certain imposition of the sentencing terms Fuelberg now faces."

On February 27, 1968, President Lyndon Johnson returned to Dallas for the first time since that fateful day nearly five years earlier. Traveling in an unmarked car at the head of a small motorcade of media and Secret Service agents, his visit shrouded in secrecy, Johnson passed within view of Dealey Plaza, where President John F. Kennedy had been shot. The president appeared to take no notice.[28]

His destination was the Dallas Memorial Auditorium, where ten thousand electric co-op leaders had gathered for the NRECA Annual Meeting. The first president to speak to the group since Eisenhower in 1950, Johnson reminisced about the birth of Pedernales Electric Cooperative. "Your vision has helped tip the balance before when the REA rescued the countryside from depression and darkness," he said. "Rural America 1968 shines with the blessings you have brought it for 30 years."

Willie Wiredhand smiled down from above and behind the president as he mentioned the theme of the meeting: "Rural-Urban Crisis: Our Challenge." "Meeting the challenge that faces us will not be easy, and the effort to revitalize rural America will necessarily involve many approaches and many groups," Johnson said. "But surely the electric cooperatives will be making one of the most valuable contributions. Because of your intimate knowledge of rural America, you are eminently qualified to exercise electric leadership." He also mentioned Vietnam. "There will be blood, sweat and tears ahead," he said. "The weak will drop from the lines, their feet sore and their voices loud."[29]

These were LBJ's people, and yet the *New York Times* noted that "he drew warmer applause for his pledge to continue assisting rural electrification than for his firmly delivered statement on Vietnam." In a few weeks, he would stun the nation by announcing that because of Vietnam, he would not run for reelection.[30]

Not long before his death, LBJ wrote a letter to John Carmody, the REA's first director. "Of all the things I have ever done, nothing has given me as much satisfaction as bringing power to the Hill Country of Texas. Today in my home county, we have full grown men who have never seen a kerosene lamp except possibly in a movie—and that is all to the good."

Despite his pioneering work for electric co-ops, LBJ faded from the pantheon of rural electric heroes—Roosevelt, Rayburn, Norris, Ellis, and others. "Year after year, other leaders of the rural electric program received NRECA's Distinguished Service Award, the highest honor bestowed on an elected official," Ted Case writes in his book *Power Plays*. "LBJ was passed over, or worse, forgotten."

NRECA CEO Glenn English told Case that Johnson was "short-changed" because of Vietnam. English took it upon himself to remedy the oversight. He made an executive decision to posthumously present the award to LBJ. In 2002, co-op leaders returned to Dallas for the sixtieth NRECA Annual Meeting. Luci Johnson, standing in for her eighty-nine-year-old mother, Lady Bird, accepted the Distinguished Service Award for her late father, gone for nearly thirty years. She read a letter from her mother in which she recalled standing with an elderly farm woman who was reaching up to switch on the light for the first time in her Hill Country home. Lady Bird, echoing her husband's letter to Carmody, wrote that such moments "gave him more satisfaction than almost anything he accomplished over his years of public service."[31]

In 2011, Pedernales Electric Co-op unveiled a Texas Historical Marker honoring LBJ's devotion to the Hill Country. "LBJ was like a force of nature," Mike Williams said at dedication ceremonies near PEC's Johnson City headquarters. "He knew what it was like to live without the quality of life that electricity eventually brought to rural America. He was determined to bring light to farms, ranches and small towns all across this country."

Many of those gathered for the marker dedication no doubt realized that the electric co-op that owed so much to LBJ was just emerging after several dark decades into the light of openness. Accompanying its new beginning was a renewed dedication to the ideals that drove the man they honored on that day. Pedernales prepared to lead as the state's electric co-ops faced new challenges in a new century.

"I HAVE SEEN MANY GREAT CHANGES, TOO MANY TO MENTION"

On a December night in 1876, Bandera County pioneer Joseph B. Hudspeth was sitting by the fireplace in the cabin he and his wife had built in Hondo Canyon. Everyone else having gone to bed, Hudspeth was savoring the quiet when suddenly his dogs outside began furiously barking. Stepping out into the moonlight, he noticed a blanket on the ground not far from the front door. Thinking the children had left it out, he strolled over to retrieve it, but when he reached down, the blanket suddenly rose up and began to scurry away into the darkness.

Underneath the blanket was an Indian. Hudspeth chased after and grabbed him, and as the two of them grappled, he yelled for his wife to come quick with his gun. Seeing the struggle, she hurried out, cocked the gun, and placed the muzzle next to the Indian's head. She pulled the trigger. The gun misfired.

Hudspeth overpowered the unexpected visitor and got him into the house, where he and his wife discovered he was a boy of about thirteen, "very active and strong," he recalled years later. He wore only a breechclout and the blanket.

The next day Hudspeth took the boy into Bandera, where he and townspeople managed to determine that he was a Tuscalero [Tuscarora] Indian who had been kidnapped by Comanches when he was about six. He had been on a raid with the Comanches the night Hudspeth found him. He had gotten separated from the raiding party and was trying to steal a horse so he could make his way back to the tribe.

"The young Tuscalero was turned over to Polly Rodriguez, a well-known guide and trailer for the Rangers," local historian J. Marvin Hunter reported in 1922. "He remained with Rodriguez many years and was known to all of our early settlers."[1]

Unfortunately, Hunter didn't record the Indian's name, although he likely would have been known to Annie Brown, another early Bandera resident. Born in Thibodeauxville, Louisiana, in 1838, Mrs. Brown lived mostly in South Texas, where, at various times during her long life, she kept a boardinghouse, worked as a freighter, raised cotton on a South Texas farm, and for thirty years worked as a nurse, part of that time on Samuel Maverick's ranch. "I live alone from choice, that I may feel free to work when I please, play or read whenever I wish, and do as I like," she explained in a 1922 oral history.

In Hunter's *Pioneer History of Bandera County: Seventy-Five Years of Intrepid History*, Brown summarized the changes she had seen:

> I have seen Southwest Texas and Bandera County change from a wilderness to a land of cultured homes; have seen the prairie schooner replaced by the automobile; have lived through the Mexican War, the Civil War, the Spanish-American War, the Philippine War, and the World War, and I hope there will never be another.
>
> The pioneer homes here had but few comforts, no luxuries. Their beds were made by driving stakes in the ground and placing split rails across; on this was placed a shuck or feather bed. The women sewed by hand, but I was fortunate in having a sewing machine. We cooked in the open fire-place. I have seen many great changes, too many to mention. I am now 83 years young, and I believe I have lived in one of the world's most interesting periods. Through it all I can see the work of an All-Wise, All-Powerful Creator. I am content.[2]

Bandera and Bandera County had changed, indeed, over the years, and yet more than a century after Annie Brown's reminiscence, she or Joseph B. Hudspeth or any of their pioneering neighbors would still have been able to find their way around. Bandera County was still ranching country, home to numerous dude ranches as well as working cattle operations, and the town was the self-proclaimed "Cowboy Capital of Texas." The sturdy limestone buildings on the courthouse square, the stately Bandera Courthouse itself, the cypress trees lining the Medina, not to mention horses tied to hitching rails downtown, would have been familiar to the pioneers. Bandera was a Texas town that had preserved its history and took pride in it.

A hundred years after Mrs. Brown's oral history, Bandera not only looked like the frontier but was also enthusiastically embracing a technological frontier. In 2017, Bandera Electric Cooperative began offering broadband service to its twenty-five thousand members across seven Hill Country counties.

"Bandera might be only an hour from San Antonio, the seventh-largest city in the United States, but that was little consolation for the town's schoolchildren," Jeff Siegel wrote in *Texas Co-op Power*. "You could see them waiting in line at the library almost every afternoon, because that was the only place in town where the computers had high-speed internet access."

Bandera CEO William Hetherington realized the co-op had an opportunity to help the community. As part of what the co-op was calling "Reimagining Rural America," Bandera began offering access to reliable high-speed broadband service delivered through fiber-optic cables, and students again were doing homework around kitchen tables or in their rooms. Parents were working from home, an indispensable service during the COVID pandemic. Families were able to take advantage of entertainment, internet shopping, and every other online offering their fellow Texans in the big cities enjoyed.

As Siegel pointed out, internet access in rural areas was not unlike access to electricity back in the 1930s, before cooperatives electrified the countryside. In 1935, when only 11 percent of US farms had electricity, President Roosevelt's REA connected rural America to the rest of the nation. Broadband began making a similar connection in the early 2000s.

A similar roadblock also existed. Like the investor-owned utilities of yore, cable and telephone companies that dominated broadband delivery saw no reason to bring the internet to areas where they couldn't be guaranteed a return on their investment. The cable and phone companies were in business to make money for shareholders; nonprofit co-ops, existing to provide service to their members, again had a vital role to play.

In the '30s, as Siegel noted, "Investors balked and claimed the infrastructure costs—building substations, placing poles and stringing wire—were prohibitive in areas with great distances containing few people. Substitute the idea of building high-speed internet's infrastructure—laying cable and building massive servers—and the result is the same."[3]

TEC head Mike Williams noted another similarity, as he explained to Texas Public Radio's Paul Flahive in a 2021 interview. Co-ops at the time were powering homes for four million Texans but were providing only around thirty thousand with broadband—fewer than 1 percent, he pointed out. Members were older and more fiscally conservative, and a number were lower income; they were as skeptical of newfangled devices as the young LBJ's Hill Country neighbors had been eighty years earlier. "They're not huge risk-takers," Williams said. "They're gonna make sure this makes sense before they do it."[4]

Bandera EC pressed ahead. In 2020, the Federal Communications Commission held a reverse auction to fund rural broadband development, awarding nearly $1.6 billion to 185 electric cooperatives around the country. Most of them competed for funding as consortiums; Bandera competed individually and became one of five co-ops nationwide to win individual funding. The co-op used the award, $1.7 million, to deploy high-speed fiber-optic broadband service to 534 additional meters.[5]

In a state where as many as one in three adults lacked access to broadband in the second decade of the twenty-first century, most living beyond the big cities, the need existed far beyond Bandera and southwest Texas. Victoria started offering broadband after polling members in 2017 about potential new services. Infinium, the co-op's broadband internet service, launched with fiber and wireless solutions for Victoria EC members and nonmembers two years later. In the spring of 2020, Dorothy Kallmayer of Port O'Connor became the one-thousandth member connected to the service. The co-op provided her with free fiber-to-the-home installation, an Amazon tablet, and other goodies.[6]

In Navasota, the livestock auction's weekly cattle sales were being broadcast all over the world thanks to MidSouth Fiber Internet from MidSouth EC. The auction was the co-op's first commercial gigabit internet customer. Bidders were witnesses to the sights and sounds of a southeast Texas cattle auction. All they missed was the camaraderie, not to mention that familiar cattle-auction odor.

"I'm more about cows than Internet, but if you were watching online, it was like you were here," Navasota Livestock Auction owner Greg Goudeau said of the first live-streamed auction. "We had people bidding live from the Panhandle to West Texas, with over 300 bids just taken on the internet. Out of 3,900 head of cattle, 1,300 was sold online."[7]

When Burleson-based United Cooperative Services voted in 2018 to provide high-speed internet service to its nearly sixty-one thousand North

Texas members, 91 percent responded positively to a ballot measure gauging interest. The proposition drew one of the largest responses for a ballot measure in the co-op's history.[8]

In early 2020, the nation's first electric co-op made its first broadband connection. Randy and Melissa Rafay became the first members of Bartlett EC to receive high-speed internet services from BEC Communications, a division of the co-op. "Over 85 years ago, rural Texans wanted access to the same conveniences afforded to those living in the cities, and electricity was at the top of the list," Bartlett general manager and CEO Bryan Lightfoot recalled. "Therefore, in 1935 our members demonstrated vision and courage when they started Bartlett Electric Cooperative, the first rural electric cooperative in the nation to provide electricity under the REA program."

BEC Communications was dedicated to providing members of the co-op in fast-growing Bell, Milam, and Williamson Counties with fixed wireless and fiber-optic internet connections. The Rafays had a wireless connection; the co-op connected its first fiber customer a week after them, Lightfoot said. "Today, lack of access to high-speed internet is a struggle our membership is facing," he continued. "But just like in 1935, the members of Bartlett EC have once again taken matters into their own hands. The launch of BEC Communications will serve to improve the lives of all our members by bringing high-speed internet to rural Texans."[9]

In his *Texas Co-op Power* article about Bandera, Siegel noted that in rural areas with broadband service, the cost of both a residential connection and a business account could be twice as high as in urban areas. Nationally, about six in ten city dwellers had access to three or more high-speed internet providers; only one in five rural residents enjoyed a choice. That disparity explained why only 55 percent of people living in rural areas had access to the speeds that qualified as broadband, while 94 percent of the urban population had access.

Reliable high-speed internet was essential if a small town like Bandera or Marlin or Paducah could ever hope to entice businesses out of the big cities. Without it, businesses simply could not function in rural areas. A call center, for example, was not a call center without high-speed internet, even if inducements like cheaper land, lower taxes, and reduced utility costs were part of the relocation package.

Broadband was essential for telemedicine, which meant that retirees who preferred to live beyond the city didn't have to pack up and drive in for a doctor's appointment. Busy parents could fit in a consultation with their children's pediatrician between soccer practice and getting dinner on the table. High-speed internet also allowed doctors to connect with each other and to medical facilities around the state or nation.

High-speed internet was vital for students in rural schools whose world suddenly broadened when they connected to university offerings, museums, and libraries far beyond their own community. In the tiny West Texas ranching town of Valentine, for example, where K–12 school enrollment was thirty-eight total in 2020, high-speed internet overcame the isolation. Valentine's junior class maintains a tradition of taco sales and other money-raising efforts for their senior trip, to such destinations as Hawaii and Australia. High-speed internet enhanced their globe-trotting adventures.

"You're not doing this [broadband] to make money. You're doing this to allow your communities to survive and to be here 20 years from now," Bandera's Hetherington told Paul Flahive of Texas Public Radio. TEC's Williams echoed Hetherington: "I frequently say to groups: 'You can't sell or provide electricity to people who aren't there anymore.'"

Broadband was perhaps the most obvious example of Texas electric cooperatives powering beyond their original mission. There were others, as the electric co-op movement looked toward a one-hundredth anniversary celebration.

In 2020, Pedernales EC became the first Texas co-op to install a 2.25-megawatt/4.5-megawatt-hour battery system. The batteries held enough electricity to power approximately two hundred homes on a hot summer day. At the time, Pedernales was serving about 340,000 active accounts, or nearly one million people.[10] "Adding new technology to our system not only helps meet the energy demands of our membership, but also supports the statewide electric grid," Pedernales general manager Julie Persley told *RE Magazine*. "We're proud to be a leader in this venture."

The battery system, which pulled power to charge and then pushed power to the grid, was expected to serve as a physical hedge to PEC's fast-growing residential load, a load that increased when the COVID pandemic drew people to the suburbs and vacation areas that Pedernales served. Charging the battery with low-cost wind energy lowered electric rates, not only for Pedernales members but also for other Texans, residents of a state with more wind than any other.

The state's electric co-ops got serious about security. In the early days, security to a co-op member likely meant an outdoor light near the henhouse to discourage an intrusive fox, perhaps an electric fence around a pasture. Nearly a century later, co-ops and other power providers were relying on control software and digital communications platforms that were growing more sophisticated at an exponential rate. As the grid grew more complex, grid assets became more attractive targets for cybercrime or cyberterrorism.

Travis Cleek, a solutions officer with a cybersecurity company called SkyHelm (and a former co-op employee), told 2021 Annual Meeting attendees that hackers spent an average of eighty-eight days inside a target's system and that 56 percent of utilities would experience an "operational downtime event" at some point. And yes, Texas electric co-ops had been attacked. "If you leave this talk and go, 'Oh, I just need to do X, and I'm protected,' then I've failed today," Cleek said. "There is not one silver bullet. Cybersecurity requires a holistic approach. Gone are the days where you have one cybersecurity person; cybersecurity is now everybody's job."[11]

Electric cooperatives had already started working on a range of defenses, from cyber-tools to continuously improving cyber-hygiene. A tool NRECA created called Essence helped co-ops detect anomalies in their network traffic in real time. With the aid of a Department of Energy grant, NRECA also developed RC3 (the Rural Cooperative Cybersecurity Capabilities Program) to offer resources and training to help co-ops improve their cybersecurity. "Electric co-ops have been using analytical tools for years, but those tools are growing in capability and are spreading," an NRECA analytics program manager told *RE Magazine*.

In addition to cybersecurity, electric co-ops in the second decade of the twenty-first century were working on technology that would quickly become commonplace, including solar farms and green energy, on-site and dispersed generation, networked microgrids, and grid resilience in response to natural disasters, terrorism, or cyberattacks.

In sun-splashed Texas, solar energy anticipated a bright future. In October 2020, the Department of Energy awarded $1 million to NRECA for research into ways to make solar energy more accessible for low- and moderate-income consumers and communities. NRECA's three-year program, Achieving Cooperative Community Equitable Solar Sources (ACCESS), explored financing mechanisms, program designs, and engagement strategies to equip electric cooperatives with the tools they needed to successfully develop solar projects to benefit consumers. "Solar energy is an integral part of the electric cooperative fuel mix, but it remains out of reach for many low-income consumers," said Jim Spiers, senior vice president of Business and Technology Strategies at NRECA. "We're excited to work towards solutions that address this challenge, particularly because electric co-ops often serve higher percentages of low- and moderate-income consumers, including 92 percent of the nation's persistent poverty counties."

The future—the immediate future, in fact—was moving at electronic speed toward energy production and delivery systems that, in the words of NRECA's Compass for the 21st Century, "are wringing out inefficiencies and driving innovation." Energy efficiency was becoming a generation alternative. Consumer and community resiliency against disruption from a variety of sources was becoming what the NRECA "Compass" labeled "the fourth public service requirement for electric providers, in addition to safety, affordability and reliability."

- In East Texas, Sam Houston EC opened an electric-vehicle charging station at the co-op's Livingston office, the only public charger within its service area.
- Bandera installed a dual-port EV charger within its service area.
- Bluebonnet EC installed 292 solar panels on its service center between Lockhart and San Marcos. The array produced enough power to offset about 70 percent of the facility's total annual energy consumption.
- Five Texas electric co-ops joined forces to plan, develop, and construct community solar projects. All seven of the solar plants—at Bartlett, Heart of Texas, and South Plains ECs and at Comanche Electric Cooperative Association (CECA) and PenTex Energy—were churning out electricity within months.
- Western Farmers EC announced that the first phase of its Skeleton Creek Project began supplying up to 250 megawatts of wind energy to the generation and transmission co-op's twenty-one member co-ops in Kansas, New Mexico, Oklahoma, and Texas. When completed, it would be the largest colocated wind, solar, and energy storage project in the United States.
- In far West Texas, Rio Grande EC was helping supply power to Blue Origin, Jeff Bezos's space exploration project near Van Horn, while in South Texas, Magic Valley EC was providing wind energy to SpaceX, the Elon Musk project at the mouth of the Rio Grande. Magic Valley, a co-op organized in 1937 to help Valley farmers grow juicy oranges and ruby-red grapefruit, was now involved with bigger orbs, namely the moon and beyond.

Electric co-ops' openness to the new, combined with the community orientation that was integral to the co-op vision, would help provide an answer to the question posed by advertising guru and business strategist Roy Spence during NRECA's process of envisioning a twenty-first-century electric co-op movement. "Electric cooperatives democratized the American Dream when they turned on the lights," Spence noted. "Can you do it again?"

A devastating natural disaster in the winter of 2021 put that existential question to the test.

"WHO SET UP THIS SYSTEM, AND WHO PERPETUATED IT?"

Valentine's Day, February 14, was on a Sunday in 2021. Throughout much of Texas, the contents of a heart-shaped box of chocolates inadvertently left outside on that day would have melted into a gooey mess. A day later, candy left out would have been frozen rock hard, as a mix of below-freezing temperatures and intermittent snow and ice gripped most of the state and for nearly a week wouldn't let go. Texas wasn't prepared, either for the history-making catastrophe or for a climate that was changing so drastically that unpredictable weather extremes soon would be commonplace.

Its official name was Winter Storm Uri, although the name never caught on; those at Uri's mercy simply called it "the ice storm" or "February." When the misery finally eased, Texas began to assess the damage, although the actual extent wouldn't be fully known for months. With temperatures settling near single digits throughout much of the state, more than two out of three Texans, 69 percent, lost electricity at some point during the storm siege. They were out for an average of forty-two hours. Almost half, 49 percent, had to do without running water for an average of more than two days. Close to 4.5 million homes and businesses were without power at Uri's peak. The storm also spawned six tornadoes on Monday, February 15. Damage estimates ran as high as $295 billion. At least 246 people died in the nation's costliest winter storm on record.

In a state already coping with a distressing multiyear COVID pandemic, the stories were heartrending. Parents, trying to keep infants and children from freezing to death, broke up furniture and burned chair legs and bookshelves in their fireplaces (if they were lucky enough to have fireplaces). To flush toilets, they used melted snow they had scraped up in the yard. Stores ran out of bottled water. At least twelve Texans died of carbon monoxide poisoning as they tried to keep warm in their vehicles. When the water came back on, burst pipes flooded already-freezing homes. Some twelve million had to boil the water to ensure its safety (that is, if they had power). Nursing homes, hospitals, and dialysis centers desperately tried to keep people alive while coping with water main breaks and oxygen shortages.

Every news outlet, every neighbor, every family member had stories to tell. In Houston, Corpus Christi, and points south, temperatures plunged into the teens—inside people's houses. In Conroe, an eleven-year-old boy died in his family's freezing mobile home after playing in the snow for the first time in his young life. A ninety-five-year-old man was found dead in his frigid house in Houston's Acres Homes neighborhood. In Sugar Land, three children and their grandmother trying to keep warm by their fireplace were killed in a house fire. In Houston, a mother and her seven-year-old daughter, desperate to get warm, died from carbon monoxide poisoning after running the family car in an attached garage. In Hillsboro, eighty-seven-year-old Gloria Jones assured her daughter in Dallas she was fine. She died during the night of hypothermia inside her home, where temperatures had fallen to fifteen degrees. In San Antonio, Manuel M. Riojas, sixty-four, struggled to breathe as power outages cut off the supply from an oxygen machine he had used since a diagnosis of esophageal cancer. He died in a hospital. In Abilene, an eighty-six-year-old woman was found dead of hypothermia in her backyard, six feet from the door to her house.

When life returned to something close to normal, Texans demanded answers. Why did the state's energy system fail? Why weren't we prepared? Who's to blame? Those were the questions millions were asking as heat, light, and water returned, accompanied by astonishingly high electric bills for many Texans.

Newly elected Harris County judge Lina Hidalgo was one of the many Houstonians who lived without power for three nights. Before the storm, Hidalgo warned her county's nearly five million residents to be prepared, but neither she nor her neighbors anticipated the catastrophe about to unfold—or the lack of preparation on the part of utility companies. "It's worth asking the question: Who set up this system and who perpetuated it knowing that the right regulation was not in place?" Hidalgo told the

Associated Press. "Those questions are going to have to be asked, and I hope that changes will come. The community deserves answers."[1]

As Hidalgo's questions implied, that two-week period during the winter of 2021 was a time of extreme testing for Texas power providers, including the state's electric cooperatives. The co-ops, the municipal utilities, the investor-owned companies—all were on the front lines, both during the crisis and afterward as both individuals and their lawmakers (meeting in Austin at the time of the storm) tried to figure out what went wrong. The chair of the state's Public Utility Commission was forced to resign. The Electric Reliability Council of Texas (ERCOT), the agency that oversees the flow of electricity across much of the state, came in for its share of well-deserved blame, as did lawmakers themselves. They were forcefully reminded that a decade earlier a fierce winter storm exposed the need for energy producers to insulate and winterize their systems for extreme cold. They didn't. Neither the Texas legislature nor the state's regulatory agencies forced them to take action. Lawmakers were reminded that campaign donations from the all-powerful fossil-fuel industry may have factored into their reluctance to act.

Most Texans rarely thought about where their power originated, until it went out—and stayed out. Only the experts would have known that nearly 50 percent of the state was powered by natural gas, 20 percent by coal, 20 percent by wind and solar, and 10 percent by nuclear energy. When it was hot outside, Texas energy systems relied on safeguards to protect against extreme temperatures. When it was cold, really cold, the safeguards for power plants and the natural gas system were revealed to be dangerously inadequate.

As snow and sleet blanketed the state and as temperatures fell, much of the infrastructure for producing and delivering electricity simply froze. Several coal plants switched off because of frozen equipment on-site. Snow- and ice-encased solar panels and wind turbines stopped functioning, forcing a series of outages throughout the system. The outages switched off electricity in various parts of the state, including sections of the natural gas system that were electrically operated. More power plants shut down. Meanwhile, shivering Texans were turning up their thermostats, driving a surge in demand.

The state's power grid began to malfunction, leading to cascading failures in the system and widespread blackouts. And Texas, proudly independent, had no resilience, no place to get help. The state did not have the option of receiving energy from other states, or even from far West Texas, where El Paso weathered the storm. El Paso was on the western grid, not the Texas grid. (Several East Texas counties were on the nation's eastern grid.)

Years earlier, Texans had chosen to go it alone, smug in the awareness that, unlike most states, their state had the ability to generate enough electricity to be self-sufficient. The state's independence allowed it to avoid federal regulation. It also meant, unfortunately, that Texans were on their own when they needed help, since it's difficult to import power from neighboring states.

"So we have a water problem—freezing water, become a gas problem, become a power problem, become a bigger gas problem, become a bigger power problem, become a water problem and a humanitarian crisis," noted Michael Webber, an energy systems expert at the University of Texas at Austin.[2]

In an online article for *Texas Monthly*, Charles Blanchard, head of natural gas research at a large commodity trading firm, focused on the state's unnatural reliance on natural gas. He pointed out that 52 percent of the electricity generated in Texas in 2020 was from natural gas, versus only 39 percent in the rest of the country. "In the low temperatures, the facilities providing one third of the state's gas production 'froze off,'" Blanchard noted. "The tanks that separate oil, gas, and water in the field got blocked, choking back production volumes. Had they not frozen, those facilities would have been able to provide enough natural gas to fuel an additional 45 gigawatts of electricity generation—far greater than the shortfall, and enough to have greatly eased, if not altogether avoided, the blackout."[3]

In the ten days between February 11 and February 20, ERCOT mandated rotating outages. The experience of Bandera EC members was typical. During the rotating blackouts, they were without power for fifteen to forty-five minutes every two or three hours over a stretch of seventy hours. That was an average of nine times a day for nearly three days. After the rotating blackouts ceased, more than three thousand Bandera members still had no power, and lineworkers continued working around the clock.

Farther west, Coleman County EC, a small co-op with about nine thousand members—most of them farmers, ranchers, and small-town residents—had a similar experience. "We still control our power-line breakers manually out in the field," general manager Synda Smith explained. "Therefore, we had linemen driving between feeders and manually turning them off and back on at the appropriate times, so each member could have some warmth throughout the storm. This went on 24 hours a day for several days. We started out with 30-minute rolling outages, then as ERCOT continued to give us less power, we were forced to move it to two-hour rolling outages."[4]

During the storm and in the days and weeks following, Coleman County's Smith was doing what her counterparts around the state were doing: patiently explaining to anxious members what had happened and what needed to be done to make sure it didn't happen again. At the height of the storm, temperatures in the Coleman County service area fell to zero, while wind speeds reached one hundred miles per hour. Co-op members suffered. "CCEC has had a plan in place for rolling outages for a long time," Smith told the *Runnels County Register*. "We just did not ever think we would have to use the plan. ERCOT decides when the load is to be shed and then contacts the transmission operators who then, in turn, contact us. They tell us how much load to shed and for how long." Smith continued, "CCEC has technology that tells us how much load is being used at each substation. Some lines were left on due to critical needs, such as nursing homes and our large gas pump stations. We manually roll all of our member's power equally to hopefully give each of our customers some power to warm their home during the zero temperatures. We had some issues and outages that happened that were beyond our control during the winter storm, but since the event, we have been working on other solutions and options to keep these same problems from happening again."[5]

Smith explained to the newspaper that Coleman County avoided one other nightmare that many energy customers experienced: they received electric bills for thousands of dollars. "The customers that experienced thousands of dollars in charges were buying power from a company that buys power on the current market," she said. "When the demand went up and gas prices went to $9,000, they in turn had to pay that high price."

Coleman County EC avoided that fate, Smith explained, because it was a member of Golden Spread EC, a generation and transmission cooperative that was able to stay online every day of the storm but Monday, when frozen gas shut down the system. "When their generators are running, they are making money to offset what had to be paid for our customers' power," Smith said. "Although it did not help us when the generators were down on Monday, it helped us in the days after when the market continued to be high."

Golden Spread, the Amarillo-based G&T, was providing power to sixteen member co-ops in the Panhandle, South Plains, and Edward Plateau regions of Texas, plus the Oklahoma Panhandle, southwest Kansas, and a small area of southern Colorado. The sixteen co-ops served about 227,000 members, including about 4,700 Coleman County EC members.

"Like all other electric co-ops, CCEC cares about its members, and we always do our best within our means to take care of them," general manager Smith said. "CCEC has had a plan in place for rolling outages for years. At the time the plan was created, I'm sure none of us ever dreamed that we would ever have to use it."[6]

Smith, focused on keeping her members safe, may not have been aware at the time, but it was touch and go for Golden Spread, as well. Because of its Panhandle location, the co-op was more accustomed to winter weather than most regions of Texas, but as the storm set in, it was buffeted by forces beyond its control. For one thing, the energy markets went crazy.

Golden Spread purchased as much natural gas as its employees could find, despite the exorbitant prices the co-op was forced to pay. In February 2021, Golden Spread spent more than $189 million on natural gas, in stark contrast to $39 million in 2020—for the whole year. "In fact," the co-op noted, "Golden Spread's February 2021 natural gas expense was roughly equal to our natural gas expense from July 2017 to December 2020."

The unprecedented fuel expense forced Golden Spread to draw down credit reserves and secure additional lines of credit just to continue operations. On the power side, Golden Spread paid $413 million for its load in February 2021. Its ERCOT load represented 71 percent of that expense, despite being only 28 percent of the total load purchased for the month. Golden Spread's load cost in February 2021 was nearly equal to the cost of all its load from November 2018 to December 2020.

Golden Spread's generators managed to hedge the co-op's exposure to high power prices sporadically throughout the winter '21 event any time it was possible to locate fuel and run. Its generation plants managed to avoid freezing until some of them were shut down on February 15 for lack of fuel. The co-op reported later that it had firm natural gas transportation arrangements in place, although its pipeline system invoked force majeure—an extraordinary event that voids a contract—on the morning of February 15. This pipeline action and a general lack of natural gas prevented the co-op from securing all its firm capacity.

"Winter Storm Uri was a catastrophe across the electric industry that will take years to fully recover from," Golden Spread reported several weeks after the event. "Rest assured, Golden Spread is working with the legislature and regulators to deal with the resulting high rates. While Golden Spread is a small player in a big market, Golden Spread commits to explore all reasonable options to support improvements to the industry, if a storm like this ever happens again."[7]

Golden Spread managed to recover within a few weeks of Winter Storm Uri despite ongoing difficulties that, as the co-op noted, would take years to resolve. A sister G&T, farther south, was not so fortunate. In March 2021, the venerable Brazos Electric Cooperative, the state's oldest G&T, declared bankruptcy. The Waco-based co-op, which served sixteen distribution co-ops representing 1.5 million Texans across sixty-eight counties, accumulated $2.1 billion in bills between February 13 and 19. In contrast, Brazos EC members paid $774 million for power for all of 2020. Headaches caused by the bankruptcy spiraled outward, causing massive difficulties for its distribution co-ops around the state.

The concatenation of events—gas-fueled power plants offline, wells frozen, a nuclear power plant shut down, and iced-over wind turbines—resulted in wholesale price spikes for electricity as high as $9,000 per megawatt-hour. In theory, the high prices acted as an incentive for power producers to create electricity. They couldn't, since much of the state's production capacity was frozen over.

Blanchard, writing in *Texas Monthly* shortly after the storm, offered detail. When power plants "could not produce the contractually agreed-upon electricity to utilities and retail electric providers, they were forced onto the spot market to buy back the volumes that they were short. The February power price for ERCOT was set at about $24 per megawatt-hour. But as power plants scrambled to find electricity to 'fill their shorts,' power prices rose quickly to the $9,000 per megawatt-hour cap set by ERCOT. This started a death spiral and explains why the shortage became so expensive, so quickly, and why many Texans will see their highest-ever utility bills for February." Blanchard went on, "Power plants found themselves looking for scarce molecules of gas that producers were also bidding on. Producers trying to maintain their contractual volumes were buying gas for $200 per million British thermal units to make up for volumes they had sold for $2.50. The main platform where energy products are traded, the Intercontinental Exchange, had to add space for another digit: no gas price had ever gone higher than $200 before. Close to Dallas, prices traded for more than $1,000 per million British thermal units."[8]

Before the storm, Brazos EC was "a financially robust, stable company with a clear vision for its future and a strong A to A+ credit rating," the co-op noted in a press release. Its credit rating was higher than that of most electric co-ops—before the $2.1 billion bill.[9]

Brazos said it received "excessively high invoices" from ERCOT for collateral and for purported cost of electric service. The invoices, totaling $2.1 billion, came due within days because, as a cooperative, Brazos's costs were passed through to its members and to retail consumers served by its members. Brazos decided that "it cannot and will not foist this catastrophic financial event on its members and those consumers," the co-op said.

On February 25, Brazos told ERCOT it would not pay the $2.1 billion sum. Clifford Karnei, the co-op's executive vice president and general manager, resigned from ERCOT's board of directors. In his personal statement filed with the court, Karnei wrote, "As the month of February 2021 began, the notion that a financially stable cooperative such as Brazos Electric would end the month preparing for bankruptcy was unfathomable. Yet that changed as a direct result of the catastrophic failures that accompanied the winter storm that blanketed the state of Texas on or about February 13, 2021 and maintained its grip of historically sub-freezing temperatures for days." Karnei emphasized that Brazos was taking drastic action to protect its member cooperatives and their more than 1.5 million retail members from unaffordable electric bills. "Brazos Electric will not foist this catastrophic 'black swan' financial event onto its members and their consumers," he said, "and commenced this bankruptcy to maintain the stability and integrity of its entire electric cooperative system."

Another G&T, Rockwall-based Rayburn EC, considered bankruptcy but ultimately decided to rely on a securitization process worked out by the Texas legislature. Rayburn, provider of wholesale electric energy and transmission services to four distribution co-ops in northeast Texas—Fannin County EC, Farmers EC, Grayson-Collin EC, and Trinity Valley EC—incurred three years of power-purchase costs in just five days," general manager David Naylor told *Forbes* magazine. He said he considered the winter-storm crisis to have been more a regulatory failure than a market one. ERCOT, he said, kept the price at the astronomically high market cap for the full five days of Uri (February 13–17) when that wasn't necessary. (It's interesting to ponder what the legendary Texan for whom the co-op was named would have had to say about the winter-storm debacle.) Naylor told *Forbes* he wasn't particularly pleased with the Texas legislature's 2021 rescue package, but it was better than bankruptcy. Authorizing the securitization of Uri's costs was a way to spread the debt over many years. Even though customers wouldn't be hit with a huge one-time rate increase, they would be paying for years to come.[10]

State leaders, legislators, and regulators vowed to fix the grid. Whether they acted with prudence and foresight in the wake of Winter Storm Uri would not be known until the next storm of a similar magnitude—or the next sustained heat wave—threatened Texas. With climate change ever more unpredictable, no one could know when that might be, just that it would be.

Many of the co-ops were skeptical, noting that the potential for problems had been obvious for years. "It was not the first time the word reliability had become the antithesis of ERCOT's performance as the state's largest grid operator," Mauri Montgomery, media relations manager for United Cooperative Services, wrote two months after the storm, "nor the first instance when PUCT diligence in oversight of the agency has been questioned. But the event did set a new benchmark for failure that will likely leave a deep scar across the Texas electric industry for some time."[11]

Cameron Smallwood, general manager for United Cooperative Services, wrote an open letter to Texas lawmakers in which he outlined a number of reforms that he believed the state urgently needed so that a winter-storm disaster would never happen again. Smallwood wanted the state to require that all natural-gas gathering, transport, and delivery be weatherized; he also wanted the state to require that every generator operating in the ERCOT market be weatherized. He wanted the state of Texas to actively support all types of new generation, and he called for reforming the ERCOT energy-only market. He wanted rules clarified and communicated to all distribution utilities related to rolling outages to eliminate vulnerability among providers. He wanted distribution utilities to be paid for providing resources to the grid, and he called for a statewide communications process to better prepare electric customers for possible future events. He also called for updating state emergency plans to include utility-related events. Finally, Smallwood urged the state to provide financial support mechanisms to keep extreme market costs from being passed down to Texas electric customers.[12]

Governor Greg Abbott either fired or forced out a number of people he had appointed to oversee the grid, but the sweeping structural reforms needed to shore up the grid fell in the face of opposition from the oil and natural gas industry. Lawmakers could not bring themselves to remove a perverse incentive for producers to profit off disaster. Instead, they settled for changing up the membership of ERCOT and the PUC, and they placed their bets on a plan that would require some power plants to winterize, although no funding was appropriated. Large segments of the natural gas industry were exempted, and penalties for noncompliance were minimal. Abbott assured his fellow Texans that he and the legislature had done all that needed to be done to make sure that a catastrophe like Winter Storm Uri would never happen again. Texans had no choice but to wait and see.

Webber, the energy systems expert from UT-Austin, was as skeptical as the co-ops, noting that Texas was building its infrastructure for yesterday's weather, not tomorrow's. "The next hundred years will be different," he said. "We know this—that the weather events will be more extreme and more frequent, which means hotter and colder, wetter and drier. So we have to deal with this [the aftermath of Uri] and design for that."[13]

The old saying in Texas—"If you don't like the weather, just wait a minute"—will get ever more true in years to come. Texas electric co-op leaders knew that and were trying to prepare, but during Winter Storm Uri they also relied on the old verities. In the tradition of the founders, co-ops around the state looked out for their members and for each other.

Mark Busby, line superintendent for Bandera EC, told Dan Oko, writing in *Texas Co-op Power*, that the storm was comparable to the worst hurricanes he had seen over three decades. The Texas Hill Country looked like the North Pole. "The roads were super snowed over," he told Oko. "Then, instead of patches of black ice, we had all black ice."[14]

With his wife and dogs trying to deal with the cold at home, Cody Hansen, a Bandera EC lineworker, parked his truck on the side of a Hill Country road and started walking, using a hammer along country roads to free frozen chains and force open gates. He had to knock ice off utility poles so he could climb to the damaged wires. "It's a lot more difficult when you have to walk the lines out and try not to break an ankle."

Hansen wasn't the only one who resorted to walking. BEC foreman John Hernandez told Oko his lineworkers kept thousands of members connected to the grid, walking when their trucks couldn't make it up a hill or got stuck in slush. Hernandez, a fifteen-year veteran at BEC, worked alongside his crew. "I've been on hurricanes and ice storms, but nothing to this magnitude," he said. "We had the snow and the ice underneath it, and then we had another snow on top of that."

Working long hours in treacherous conditions, Bandera's lineworkers were often greeted with coffee and hot food when they were near a house. "We were very blessed to have members help us out if we needed it," Hernandez said. "For linemen, it's hard to go home at night knowing someone is still without power." Brad Elliott, a second-year apprentice lineworker, worked twenty-three hours straight. Like Hernandez, he was grateful when a member offered a quick snack or a cup of coffee. "I've always had a passion for doing something that was benefiting people," he told Oko. "That right there does it for me. Going every day, especially in a storm, that's when it really counts when you're getting everybody's electricity back."

It's the rare lineworker who hasn't been called out to help another co-op in the aftermath of a hurricane, tornado, wildfire, or some other disaster that wreaks destruction on a community. Despite the crisis at home, the co-ops were helping co-ops nearby. "That's what co-ops are for—to help each other," Tate Glasscock, foreman at Floydada-based Lighthouse EC, told Chris Burrows, a senior communications specialist at Texas Electric Cooperatives.

For the June 2021 issue of *Texas Co-op Power*, Burrows compiled a list of recollections from lineworkers, managers, office workers, and members pulling together through a crisis. Kathi Calvert, general manager at Houston County EC, expressed her gratitude, echoing many of her counterparts across the state. "When our system was on the ground and members were without power, it was reassuring to know we had support from our co-op family," she told Burrows. Bryan Chandler, operations manager at McGregor-based Heart of Texas EC, said, "Our crews in the field were constantly stuck and having to be pulled around due to the icy conditions. Our members came to our aid with food, fuel and tractors to help pull trucks." Dangerous roads, iced-over lines and poles, and grid-mandated outages made lineworkers' jobs even more challenging. Brad Morrow, a United EC lineworker, told Burrows that simply knowing that people were in need kept him going. Members coping with their own problems found time to lend support online and in the field. A woman in Moss Bluff, Louisiana, for example, cooked up a meal of gumbo "and any kind of Louisiana fixin's you could think of," Morrow said, for lineworkers from Texas and Missouri who restored her power after she had spent weeks living off a generator. "It's the people who make it worthwhile," Morrow said. "You'll never find someone more thankful and willing to help you when you're there to help them."

When February's polar vortex brought to Texas enough snow, ice, cold, and wind to strain every bit of infrastructure, lineworkers across the state faced fender benders and exhausting conditions as they worked to keep power flowing to co-op members. "There was the normal fatigue that comes from working so many 12-hour days in a row, but sometimes that was compounded by coming home to a cold house," said Kendal Fiebrich, a Bluebonnet EC lineworker.

"We all have families—children and babies at home and parents and grandparents whom we care for—and knowing we had no choice but to shut off power to many of you was heart-wrenching," Mark McClain, Big Country EC's general manager, explained to his members in the local pages of *Texas Co-op Power*. "We pride ourselves on restoring power as quickly as possible, but this time it wasn't a matter of replacing poles, crossarms or wire—scenarios that we could have fixed. To the greatest degree, we had no control over this, and it made us as miserable as you were."

After Texas thawed and the lights came back on in late February, lineworkers still weren't done. About a dozen co-ops sent workers to other cooperatives after ice broke thousands of poles across Co-op Country and residential heating demand overwhelmed other electric equipment, stretching co-ops and crews thin.

"This was definitely the worst winter storm I have worked in," said Doug Grimm, a twenty-year lineworker veteran for Bluebonnet EC. "We did what we always do: Come together and get the job done."

"THE GREATEST THING ON EARTH IS TO HAVE THE LOVE OF GOD IN YOUR HEART, AND THE NEXT GREATEST THING IS TO HAVE ELECTRICITY IN YOUR HOUSE"

If Texans learned anything from Winter Storm Uri in 2021, we learned that we live in perilous, unpredictable times. A pandemic, drastic climate change, carbon constraints, and global unrest—as well as unforeseen inventions and new ways of meeting the world's many challenges—all combine to humble any soothsayer.

Of course, the unforeseen is nothing new. The electric co-op movement itself was a response to peril and unpredictability. A dozen years into the twenty-first century, that unpredictability—and the concomitant need for a map, if at all possible—brought together a twelve-person committee of electric cooperative leaders from around the country. For a year, the group met periodically to consider the future of electric co-ops and their distinctive way of doing business. Initiated by the late Glenn English, former Oklahoma congressman and later NRECA CEO, and chaired by Mike Williams, longtime CEO and president of Texas Electric Cooperatives, the group called themselves the NRECA 21st Century Cooperative Committee. Their mission was to address what was, in essence, an existential question: "Does the electric cooperative model still make sense 75 years after the first rural electric cooperatives were created?"

The committee met in various places across the country. Members conferred with industry observers and consultants in Boston, Chicago, and Denver. They observed focus groups of co-op CEOs and board members and conducted consumer interviews in three cooperatives, including Burleson-based United Cooperative Services. They conferred with writers who had conducted research on the role of cooperatives in society and took a field trip to Withlacoochee River Electric Cooperative in Florida, where they explored how community enrichment functioned as an integral part of the co-op's operations. The NRECA research team also contributed time, expertise, and important perspectives on the co-op model.

The initial meeting was in Los Angeles, where the group got a clearer sense of how to frame and articulate its mission from Texas business consultants Haley Rushing and Roy Spence, coauthors in 2009 of *It's Not What You Sell, It's What You Stand For*, a book about purpose-based brand management.

San Antonio native Rushing was cofounder with Spence of the Purpose Institute, where her title was "chief purposologist." Her Austin-based company had worked with the American Red Cross, Southwest Airlines, Whole Foods Market, and other organizations seeking to sharpen their understanding of goals and objectives and crafting strategies for reaching them. She challenged her co-op clients with questions: As industries and businesses evolve, how do you ensure your purpose stays relevant? What are your core values? What is the deeper purpose of the work you do? For the next year, her co-op clients would be seeking answers to her pertinent—and unsettling—questions.

Spence, a Brownwood native, graduated from UT-Austin in 1969. Two years later he and three college buddies founded the Austin-based advertising agency GSD&M with no clients and virtually no money. For the first few months of the company's precarious existence, Spence slept on a mattress under the art table in the company's tiny one-room office, rent eighty-five dollars a month. He bathed in the health-club swimming pool next door.

By the end of the decade, GSD&M had become one of the most successful advertising agencies in the country, regularly overseeing multimillion-dollar accounts. Its early clients included Southwest Airlines, Walmart, DreamWorks, BMW, and the state of Texas. Developing a campaign in the 1980s to persuade Texas drivers to stop tossing trash out their vehicle windows, GSD&M (and Spence in particular) became famous for the slogan that turned into a catchphrase worldwide: "Don't mess with Texas."

With his West Texas twang, folksy personality, and knack for making connections between client and customer, Spence also had become a renowned motivational speaker, as well as a sought-after business consultant. Advising the nation's electric cooperatives in Los Angeles one morning in 2012, he quickly got to the heart of the issue. "Electric cooperatives democratized the American Dream when they turned on the lights," he pointed out. "Can you do it again?"[1]

Williams recalled a spirited discussion about purpose at that first meeting. "Isn't it obvious?" a few committee members quickly insisted. "Our purpose is to provide safe, reliable and affordable electricity to our members. It's that simple."

Well, yes, but as the conversation continued, they all began to realize that maybe it wasn't just "that simple." Committee members were struck with the dawning realization that, in addition to providing electric power to their members—which every power company does—they were also improving the quality of their members' lives. Safe, affordable, and reliable electricity was simply a manifestation of that noble purpose.

The co-op folks were speaking Spence's—and Rushing's—language (not to mention the language of LBJ, Clyde Ellis, and other electric co-op pioneers). Spence came by it naturally. "Fiercely proud of his rural upbringing, he's a champion for the entrepreneurial underdog," Camille Wheeler wrote in a Spence profile that appeared in the March 2012 issue of *Texas Co-op Power*. "It's a principle based on community and the undergirding of Texas that draws its strength from its rural vitality—from member-owned electric cooperatives putting people first, to farmers, ranchers and small-business owners."[2]

The redoubtable ad man challenged the co-op committee with an even more direct—and unsettling—question: "If electric cooperatives did not exist, would it matter?"[3] The answer, the committee said in a 2013 report, was a resounding "yes," but that "yes" came with conditions. The 105-page report, *The Electric Cooperative Purpose: A Compass for the 21st Century*, explored those conditions in detail.

Looking back on the early years, committee members noted that founding members in the 1930s understood that the co-op was "their" organization. They relied on their co-op to provide services they needed, beginning with electricity. Institutions like electric co-ops were just as necessary decades later, but the same sense of ownership and hope that energized the generation who first "saw the lights come on" needed to be reenergized. The committee concluded it would be just as important in the 2030s as it was in the 1930s for member-owners to believe that their cooperative "cares about me."[4]

The committee drafted a purpose statement to serve as a compass to help electric co-ops navigate the challenges of the future: "The Purpose of the electric cooperatives is to power communities and empower members to improve the quality of their lives."[5] As Rushing pointed out, that purpose should guide actions, foster innovations, inspire employees, and pave the path to high performance.

How does an electric co-op differ from an investor-owned utility? The most distinctive difference, the committee concluded, was that the two entities had different concepts of the "bottom line." The bottom line for IOUs was, of course, profit; for the electric cooperative business, the bottom line was, as the final report noted, "the empowerment of its member-owners." Of course, prudent financial management and solid finances were prerequisites for both co-ops and for-profit companies if they were to achieve the bottom lines they sought, but for the co-ops something else was essential.

"The electric cooperative story is about ordinary people who banded together to improve their quality of life by bringing electricity to their communities when no one else would," the committee noted. "Delivering a mechanism for empowering people to improve their quality of life is not only the history of electric cooperatives, it is also our heritage—it is who we are. Some might say it is the soul of the electric cooperative movement."[6]

The committee espoused community enrichment and enhancing the capacity of member-owners for self-help, objectives that were inspiring but tended to blend into hard-to-grasp abstraction. Fortunately, the group researched case studies to make its purpose-driven goal concrete.

- The Withlacoochee River Electric Cooperative near Tampa, Florida, not only provided power to more than two hundred thousand members but also created a for-profit economic development subsidiary that focused on strengthening the local economy, creating jobs (especially for younger workers), enhancing the tax base, and attracting steady commercial and industrial loads to help keep rates lower for WREC members. The co-op also created and managed two foundations, one to help co-op members cope with catastrophic events, the other to help the children of co-op members further their education. WREC's most ambitious project was to revitalize a once-thriving community that had been in decline since a lumber mill closed in 1959. The effort involved building new homes under the auspices of Habitat for Humanity, refurbishing existing homes, working with the county to repave roads and improve lighting and water, building a football field and a soccer field, and securing matching funds for a new community center.[7]

- United Power (UP) near Denver, Colorado, became aware that its members were eager to be part of green energy alternatives. After exploring various options, UP formed the nation's first rural electric cooperative solar farm, Sol Partners Cooperative. The co-op also developed a novel approach to solar power: instead of members leasing and installing someone else's solar panels atop their houses, or purchasing the expensive equipment outright, the co-op bought, installed, insured, and maintained the solar panels at a central site, a farm, in which any member could invest. Using land UP already owned, the co-op offered members the opportunity to lease a single 210-watt solar photovoltaic panel for twenty-five years for an up-front cost of $1,050. The co-op estimated that the payback could be realized within twenty to twenty-five years.[8]
- North West Rural EC, based in northwest Iowa, responding to the farm and energy crises of the 1980s, began partnering with other cooperative and community groups to increase value-added agriculture across the region. Tangible results included industrial park development and broadband, grants for a school performing-arts center, new radiology equipment for a local hospital, and two built-on-spec commercial buildings. The co-op also assisted a family-owned ice cream manufacturer to increase milk output in the region, along with assistance to various food-processing plants. As a result, the ice cream manufacturer transformed the small town of Le Mars into the self-proclaimed "ice cream capital of the world," and a number of food processors, drawn to the ice cream operation, strengthened the local economic base.[9]
- Central Wisconsin Electric Cooperative, a relatively small co-op serving about 7,100 members, "has made community involvement part of the cooperative's culture, and staff are expected to be involved," co-op CEO Greg Blum, a former Peace Corps volunteer, told the committee. CWEC was involved in starting a business park, constructing an assisted-living facility for seniors, and initiating a Capital Credits for Conservation program. The co-op also ran a scholarship and youth leadership effort and a program for members who needed help with their bills and weatherization.[10]

In 2012, Mike Williams reminded Camille Wheeler, writing in *Texas Co-op Power*, that the state's electric cooperatives were products of a history that functioned as a sturdy foundation. "Over 75 years ago, we brought light to vast areas of the country that other utilities did not want to serve," he said. "The reality is we brought more than light; we brought a quality of life to small towns and rural areas. As important as that was then, maintaining that quality of life may be even more critical today. And electric cooperatives are uniquely positioned to fill that role."[11]

With the Texas economy booming a decade later, with newcomers flooding into a once-rural state that was becoming urban and suburban at a breathtaking pace, who remembered the co-ops' proud history? Who cared?

At the 2012 Los Angeles meeting, Spence shared an old story about an elderly Tennessee farmer who was an electric co-op member in the 1940s. "The greatest thing on Earth is to have the love of God in your heart, and the next greatest thing is to have electricity in your house," the farmer proclaimed.

With Google, Amazon, Samsung, Tesla, and other huge, power-intensive companies moving into Texas and into co-op territory, was that proud history relevant? Would Jeff Bezos and Elon Musk feel the same way about their power provider as that Tennessee farmer felt?

Rayburn Country's David Naylor worried that co-ops might have difficulty "getting a seat at the table." The big boys might run over them, in other words. Naylor told *Forbes* that the newcomers might be unaware of the situation on the ground when it came to electricity supply. "This has made it hard for us to be heard," he said. The Rayburn Country CEO offered an example: the newcomers might buy up the entire electrical output of a big solar installation, enough for their nationwide operation, when they needed only 29 percent of that for their Texas operations. "One of our four distribution co-op members might want to buy some of that power," he said. He explained that the four Rayburn members were responsible for power in their service areas and that they needed to educate the power-hungry newcomers on the operation of the electric supply infrastructure in Texas. Naylor didn't regard them as competitors. Rather, they were neighbors in need of education. He welcomed them as customers, as heirs in good standing to a proud co-op tradition.[12]

The newcomers also forced co-ops to acknowledge that they had to be as agile and at the same time as focused as co-op pioneers. Not only did they have to respond to their customers' needs, as Spence and Rushing urged, but they also had to lead. Facing a precarious and uncertain climate future, co-ops had to be resilient enough to shift tactics and techniques, experiment, and create new energy paradigms.

In the years following the deadly 2021 storm, an increasing number of studies showed that Texas and the United States were simply not prepared to avoid a repeat. Co-ops specifically and the power industry in general had a lot of work still to do to build a more resilient infrastructure. They weren't prepared for the extremes that were coming.

"Climate resilience is not just a mere buzzword in policy and activism circles. It's a necessity to manage the risks associated with a warming world as well as the weather events we already face," experts told the *Washington Post*.[13] One of those experts, Rich Sorkin, founder and chief executive of Jupiter Intelligence, a company that helps governments and companies manage climate risks, was particularly dire. "It's the same dynamic whether we're talking about fire in California and Spain, heat in Dubai and Phoenix, flooding in Florida and Tokyo, cold, wind and flooding in Texas, etc., etc.," he told the *Post*. "The vast majority of these places are livable with sufficient planning and investment for quite some time. Without that planning and investment, a hellscape will be upon us."[14]

Another study contended that "a hellscape" could be avoided by a total conversion to clean and renewable solar, wind, and water energy sources—and that such a conversion was possible. Conducted by an organization called Renewable Energy, the study noted that wind, water, and solar already accounted for about a fifth of the nation's electricity, although a full transition in many areas was slow. It also contended that a switch to renewables would lower energy requirements, reduce consumer costs, create millions of new jobs, and improve people's health.[15]

As the *Post* reported, the researchers modeled grid stability throughout the contiguous United States, including data from a weather–climate–air pollution model. Using energy consumption data from the Energy Information Administration, the research team simulated energy demands for 2050 to 2051. The team found that the actual energy demand decreased significantly by simply shifting to renewable resources, since they were much more efficient. For the entire United States, total end-use energy demand decreased by about 57 percent. Per capita household annual energy costs were about 63 percent less than in a "business as usual" scenario.

In Texas, the study found, a complete green transition would reduce the annual average end-use power demand by 56 percent. It would also reduce peak loads, or the highest amount of energy drawn from the grid at a time. The study anticipated that many homes would also have their own storage and wouldn't need to rely on the grid as much. "Everything that we currently do using fossil fuels would be done using technology that is run through electricity," Anna-Katharina von Krauland, a coauthor of the Renewable Energy report, told the *Post*. "The amount of energy that's needed to perform activities, basically to turn on the light or to fuel industrial processes, that would actually be decreased if you use more efficient energy supply."[16]

In December 2020, *RE Magazine* peered into the future in an effort to anticipate electric co-ops in the year 2045. Interviewing a panel of industry experts and futurists, the one area of consensus was the inevitably of surprise. "There are companies that do not exist that will build products not yet conceived using materials not yet discovered with methodologies not yet invented," Jim Carroll, a futurist and innovations expert, told the magazine.[17]

With due acknowledgment of the future's unpredictability, the experts weren't shy about anticipating the world in which tomorrow's electric co-ops will be operating:

- Renewables will be the fastest-growing source of electricity through 2050, in conjunction with evolving battery technologies.
- DIY (do it yourself) will appeal to more and more Americans, particularly as solar and batteries become a greater part of everyday life.
- Concern about climate change will drive a continued shift away from carbon-based fuels and toward electricity. Electric vehicles, for example, will account for 61 percent of all US passenger vehicle sales by 2040.
- Global warming will bring a dramatic shift in migration patterns, with more people moving inland and away from the coasts, as well as toward rural areas.

"Whether it's fusion, advances in fuel cells, battery technology, or something completely unknown today, experts agree that the next quarter-century will be one of unprecedented change and will transform the power industry and electric cooperatives," *RE* concluded.

Change is always a challenge, as Jill Cliburn, president of a New Mexico–based consultancy called Cliburn and Associates, noted in 2022. "Electric co-ops are in the heartland of America . . . making communities work where you may not have ready access to everything you get in a big city."[18]

Working with co-ops to use more solar energy and battery storage, Cliburn's team created online resources that introduced electric co-ops to how the clean energy industry works and then served as a guide through the initial steps of planning and purchasing. The co-ops "have boards of directors, which are very capable people in the community, but they come from a variety of backgrounds," Cliburn said. "So they may not be engineers. They may be somebody from the school district, somebody from a private business."

Anyone aware of electric co-op history would find such challenges familiar. These are the same rural and small-town folks who, in the early years, fought the giant private utilities to a standstill. As the pace of change continued to accelerate in the early decades of the twenty-first century, once-rural electric co-ops seemed up to the challenge. The larger ones in particular, those facing big-city and big-industry demands, were arguably more nimble than their for-profit counterparts. They were accustomed to adapting to member needs, as they acknowledged in Los Angeles in 2012.

Decades after their founding, they were still serving rural Texans, to be sure, but they were also helping meet the power needs of Jeff Bezos's audacious outer-space probes in far West Texas and those of Elon Musk at the mouth of the Rio Grande. They were still providing light and power to farms and ranches across a vast state, even as they continued to empower some of the fastest-growing suburbs in the nation. And Mike Williams, ten years after the "Compass for the 21st Century" assessment, was still serving Texas Electric Cooperatives, still insisting in his role as CEO that the co-ops' focus on the needs of their members offered a distinct advantage, whatever the future brought. "I do believe that the conclusions [of the NRECA committee] are still relevant today," Williams said. "I think the most significant contribution was to acknowledge that our real purpose was and is and likely will be to improve our members' lives."[19]

Nearly a century after co-ops brought light and power and a better way of life to rural Texas, the essential question that Roy Spence asked redounded through the years: "Can you do it again?"

Busy, service-oriented, innovative electric co-ops across Texas had an answer to Spence's challenging question. The answer was "Yes!" Without a doubt.

NOTES

Preface

1. Renee D. Cross, Mark P. Jones, Pablo Pinto, and Kirk Watson, *The Winter Storm of 2021* (Houston: Hobby School of Public Affairs, University of Houston, 2021).

2. Renee D. Cross, Mark P. Jones, Pablo Pinto, and Kirk Watson, *Local Electricity Utility Provider Performance during the 2021 Winter Storm* (Houston: Hobby School of Public Affairs, University of Houston, 2021), 3.

3. Ibid.

4. Joe Holley, "Co-ops Shine," *Texas Co-op Power*, October 2021, 13.

5. Ibid.

6. Ibid.

7. Ibid.

8. Ibid., 14.

9. Ibid.

10. Synda Smith, phone interview by the author, February 25, 2022, Coleman, TX.

11. Holley, "Co-ops Shine," 14.

12. Ibid., 13.

13. Robert Caro, *The Path to Power*, vol. 1 of *The Years of Lyndon Johnson* (New York: Alfred A. Knopf, 1983), 502–15.

14. Marquis Childs, *The Farmer Takes a Hand: The Electric Power Revolution in Rural America* (Garden City, NY: Doubleday, 1952), 13.

15. Ibid.

16. Bill Minutaglio, interview by the author, May 13, 2022, Austin, TX.

17. Robert J. Duncan, "Lowman, Albert Terry [Al]," *Handbook of Texas Online*, Texas State Historical Association, accessed May 14, 2022, https://www.tshaonline.org/handbook/entries/lowman-albert-terry-al

18. Sarah Greene, "'Moment of Magic' Is Ex-Boy's Treasured Memory," *Gilmer Mirror*, Special Upshur Rural Section, November 4, 1987.

Chapter 1

1. "Laying Corner Stone," *Austin Daily Statesman*, July 5, 1885, 1.

2. Ibid.

3. Ibid.
4. Edmund Morris, *Edison* (New York: Random House, 2019), 583.
5. Ibid., 588.
6. Ibid., 597.
7. Edison Museum, Beaumont, TX, https://www.edisonmuseum.org.
8. Richard M. Hausler, "The Story of Electricity—a Key and a Kite Turned On the Lights," *Texas Co-op Power*, January 1951, 3.
9. Joshua C. Gregory, "The Life and Work of Sir Humphry Davy," *Science Progress in the Twentieth Century (1919–1933)* 24, no. 95 (January 1930): 485–98.
10. Morris, *Edison*, 521.
11. Ernest Freeberg, *The Age of Edison: Electric Light and the Invention of Modern America* (New York: Penguin Press, 2013), 69.
12. Morris, *Edison*, 384.
13. Michael Lind, *Land of Promise: An Economic History of the United States* (New York: Harper, 2012), 195.
14. "Edison's Electric Light: 'The Times' Building Illuminated by Electricity," *New York Times*, September 5, 1882, 8.
15. Morris, *Edison*, 425.
16. "Palace of a Magician," *Austin Daily Statesman*, May 23, 1898.
17. Ibid.
18. Ibid.
19. Ibid.
20. Ibid.
21. Casey Greene, Rosenberg Scholar, Rosenberg Library, email query by the author, August 21, 2021, Galveston, TX.
22. Robin Raleigh, "When Electricity Came to Dallas," Dallas Gateway, January 23, 2018, accessed August 3, 2021, https://dallasgateway.com (cited post has since been removed from website).
23. Ibid.
24. Ibid.
25. Ibid.
26. William Owens, *This Stubborn Soil* (London: Faber and Faber, 1966), 240.
27. "Fiat Lux!," *Waco Examiner*, September 19, 1885, 1.
28. Leanna Barcelona, "When Did the Lights First Shine Bright?," April 23, 2018, Texas Collection, Baylor University, Waco, TX.
29. "Our History: From Mule Drawn Streetcars to Solar Energy," El Paso Electric, https://www.epelectric.com/company/about-epe/history/from-mule-drawn-streetcars-to-solar-energy
30. Ibid.
31. Ibid.
32. Ibid.
33. Ibid.
34. Ibid.
35. *Brenham Weekly Banner*, November 11, 1897.
36. *Abilene Reporter*, February 17, 1893, 11.

37. *Galveston Daily News*, February 23, 1878.
38. *Fort Worth Daily Gazette*, February 24, 1884.
39. Caro, *Path to Power*, 510.
40. Ibid.
41. Ibid., 511.

Chapter 2

1. Caro, *Path to Power*, 307.
2. Joe Holley, "'Mr. Sam' Couldn't Be Bought,'" *Houston Chronicle*, March 10, 2016.
3. Ibid.
4. D. B. Hardeman and Donald C. Bacon, *Rayburn: A Biography* (Austin: Texas Monthly Press, 1987), 35.
5. Ibid., 36.
6. Ibid.
7. Ibid.
8. Caro, *Path to Power*, 310.
9. Hardeman and Bacon, *Rayburn*, 47.
10. Ibid., 48.
11. Neal Johnson, *And There Was Light: The History of Kaufman County Electric Cooperative, 1937–1997*, in collaboration with Ray Raymond (Austin: Texas Electric Cooperatives, 1998), 11.
12. Hardeman and Bacon, *Rayburn*, 145.
13. Ibid., 148.
14. Ibid., 170–71.
15. Studs Terkel, *Hard Times: An Oral History of the Great Depression* (New York: Pantheon Books, 1970), 271.
16. Hardeman and Bacon, *Rayburn*, 174.
17. Ibid., 180.
18. Ibid., 172.
19. Bob Poage, oral history, 523, Texas Collection, Baylor University Institute for Oral History, Waco, TX.
20. Hardeman and Bacon, *Rayburn*, 174.
21. Ibid., 197.
22. Ibid., 199.

Chapter 3

1. Richard A. Pence, ed., *The Next Greatest Thing: 50 Years of Rural Electrification in America*, (Washington, DC: National Rural Electric Cooperative Association, 1984), 39.
2. Childs, *Farmer Takes a Hand*, 38.
3. Ibid., 40.

4. "Electric Farm Tests Approved by 'U' Regents," *Minneapolis Daily Star*, April 16, 1924, 1.
5. "Electricity and the Farmer," *Minneapolis Daily Star*, October 10, 1924, 14.
6. Childs, *Farmer Takes a Hand*, 40.
7. Caro, *Path to Power*, 516.
8. Ibid.
9. Childs, *Farmer Takes a Hand*, 42.
10. Caro, *Path to Power*, 517.
11. Ibid., 518.
12. Childs, *Farmer Takes a Hand*, 48.
13. Pence, *Next Greatest Thing*, 44.
14. Ibid.
15. Ibid., 59–60.
16. Ibid., 61–62.
17. Ibid., 62.
18. Ibid.
19. Ibid., 63.
20. Noah Karr Kaitin, "How It Worked When It Worked: Electrifying Rural America" (senior honors thesis, Cornell University, College of Industrial and Labor Relations, 2013), 11.
21. Childs, *Farmer Takes a Hand*, 51.
22. Pence, *Next Greatest Thing*, 65.
23. Ibid.
24. H. S. Person, "The Rural Electrification Administration in Perspective," *Agricultural History*, April 1950, 73.
25. Childs, *Farmer Takes a Hand*, 56.
26. Ibid., 62.
27. Ibid.
28. David J. Thompson, *Weavers of Dreams: Founders of the Modern Co-operative Movement* (Davis: Regents of the University of California, 1994), 3–4.
29. Ibid., 5.
30. Ibid., 4.
31. Ibid., 34.
32. Ibid., 35.
33. David J. Thompson, *Weavers of Dreams: Founders of the Modern Co-operative Movement* (Davis: Regents of the University of California, 1994), 46, 48, accessed May 28, 2023, ucawr.edu/sites/sfp/files/143837.pdf
34. Ibid.
35. Ibid.
36. "100th Anniversary of the Birth of the Cooperative Movement," *Texas Co-op Power*, October 1944, 8.
37. Robert Gibson, "Fate of First Modern Co-op Reflects Changes," *Texas Co-op Power*, October 1985, 4.
38. Thompson, *Weavers of Dreams*, 142.
39. Pence, *Next Greatest Thing*, 81.

40. Person, "Rural Electrification Administration," 73.

41. Hardeman and Bacon, *Rayburn*, 201.

42. Wright Patman, interview by Joe B. Frantz, August 11, 1972, interview 1 (I), oral history transcript, 8, LBJ Library Oral Histories, LBJ Presidential Library, accessed May 20, 2022, https://www.discoverlbj.org/item/oh-patmanw-19720811-1-74-97

43. Bob Poage, oral history, 774, Texas Collection, Baylor University Institute for Oral History, Waco, TX.

44. Hardeman and Bacon, *Rayburn*, 203.

45. Ibid.

46. Childs, *Farmer Takes a Hand*, 70.

Chapter 4

1. Carol Moczygemba, "Polishing Up the 'Gem City of the Blackland Belt,'" *Texas Co-op Power*, October 1998, 6.

2. Clay Coppedge, "The Old Bartlett and Western Railroad and Marie Cronin," Letters from Central Texas column, *Texas Escapes Online Magazine*, April 27, 2006, accessed February 14, 2021, texasescapes.com.

3. Jack E. Nettles, "Bartlett, First in the United States," *Texas Co-op Power*, February 1951, 7.

4. Ibid.

5. "First REA-Financed Project in Nation Was Started near Bartlett, Texas, 9 Years Ago," *Texas Co-op Power*, March 1945, 7.

6. Ibid.

7. Charles Boisseau, "Historic Connection," *Texas Co-op Power*, October 2010, 6.

8. Johnson, *And There Was Light*, 11.

9. Ibid., 1.

10. Ibid., 18.

11. Ibid.

12. Ibid., 21.

13. "Swisher EC Celebrates Rich 75-Year History," *Swisher Electric Cooperative 2013 Annual Report*, 19.

14. Ibid.

15. Ibid.

16. Patricia Ward Wallace, *Waco: A Sesquicentennial History* (Virginia Beach, VA: Donning, 1999), 110.

17. David Frost, interview by the author, June 28, 2021, Douglassville, TX.

18. Robert Kerr, "Miss Mabel," *Texarkana Gazette*, 1982, reprinted in *Texas Co-op Power*, October 1983.

19. Warren Burkett, "The Livestock Auction That Started a Co-op," *Texas Co-op Power*, February 1961, 9.

20. Ibid.

21. Ibid.

22. Ibid.
23. Ibid.
24. Cecil Harper Jr., "Bowie County," *Handbook of Texas Online*, Texas State Historical Association, accessed May 19, 2022, https://www.tshaonline.org/handbook/entries/bowie-county
25. Tom Frost, interview by the author, July 1, 2021, Douglassville, TX.
26. Burkett, "Livestock Auction," 9.
27. "Our History," Bowie-Cass Electric Co-operative, accessed July 15, 2021, bcec.com.
28. Kerr, "Miss Mabel," 7.
29. Tom Frost, interview.
30. "Our History," Bowie-Cass Electric Co-op.
31. Kerr, "Miss Mabel," 7.

Chapter 5

1. Ronnie Dugger, *The Politician: The Drive for Power, from the Frontier to Master of the Senate* (New York: W. W. Norton, 1982), 208.
2. Ibid., 209.
3. Ibid.
4. Ibid.
5. "Federal Agency Aids Power Lines; Lower Colorado River Authority Contract Reveals First Evidence of Coordination," *New York Times*, July 25, 1939, cited in Dugger, *Politician*.
6. Dugger, *Politician*, 209.
7. Lawrence Goodwyn, *Democratic Promise: The Populist Moment in America* (New York: Oxford University Press, 1976), 109–53.
8. Tom Wicker, "With Johnson on the Ranch; the President, Observed at Home, Reveals Image of a Westerner," *New York Times*, January 5, 1964.
9. Caro, *Path to Power*, 524.
10. Ibid., 524–25.
11. Jack E. Nettles, "Jackson Electric Co-op Brings 'Face-Lifting' to Gulf Coast Area," *Texas Co-op Power*, April 1952, 8.
12. Caro, *Path to Power*, 523.
13. Ibid.
14. Ibid., 524.
15. Ibid.
16. Ibid., 522.
17. Elizabeth Carpenter, "From Coal Oil to Kilowatts," *Corpus Christi Caller-Times*, August 20, 1950, 59.
18. Caro, *Path to Power*, 502.
19. Ibid., 527.
20. Dugger, *Politician*, 212.
21. Ibid.
22. Caro, *Path to Power*, 527.

23. Dugger, *Politician*, 213–14.
24. Ibid.
25. Carpenter, "From Coal Oil to Kilowatts."
26. Walter Richter, "Government Can Be a Force for Improving Society," *Texas Co-op Power*, August 1982, 4.
27. Sim Gideon, interview by Paul Bolton, October 3, 1968, oral history transcript, Discover LBJ, LBJ Presidential Library, accessed August 26, 2021, https://www.lbjlibrary.org
28. Ed Crowell, *Seventy-Five Years of Power: The Story of Bluebonnet Electric Cooperative* (Bastrop, TX: Bluebonnet Electric Cooperative, 2014).
29. Ibid.
30. Clayton Stromberger, "The Power of Memories," *Texas Co-op Power*, November 2019, 25.
31. Clayton Stromberger, "A Waltz through Time," *Texas Co-op Power*, September 2019, 22, 2.
32. Brian K. Moreland, *Dreams Come True: The 75 Year History of Hilco Electric Cooperative, Inc., 1937–2012* (Itasca, TX: Hilco Electric Cooperative, 2010), 7.
33. Ibid., 5, 9, 19–20, 22–23.
34. Childs, *Farmer Takes a Hand*, 118.
35. Ibid., 72.

Chapter 6

1. "Zabcikville," Texas Historical Marker, intersection of Airville Road and State Highway 53, Temple, TX.
2. "Czech Immigrants to Bell County" (folder), Bell County Museum, Belton, TX.
3. Ibid.
4. Ibid.
5. "How Can We Get Electric Power?," *Texas Co-op Power*, May 1957, 10.
6. Ibid.
7. Childs, *Farmer Takes a Hand*, 78–79.
8. William S. Roberts, "'Pirating' of Members Facing Co-ops in 15 States," *Texas Co-op Power*, April 1953, 2.
9. Ibid.
10. Ibid.
11. Ibid.
12. "How Can We Get Electric Power?," 10.
13. "Belfalls Outlasts 'Claim Jumpers,'" *Texas Co-op Power*, June 1951, 11.
14. "How Can We Get Electric Power?," 10.
15. George W. Haggard, "Between the Lines/Editorial Comment," *Texas Cooperative Electrical Power*, September 1944, 2.

Chapter 7

1. "Hand Operated Sheller Converted to Motor Use Proves Great Success," *Texas Cooperative Electric Power*, July 1944, 2.
2. "Small-Scale Dairying Earns Big Profits," *Texas Co-op Power*, November 1947, 3.
3. "Fayette Farmer Finds Electric Pump Beats 'Old Oaken Bucket,'" *Texas Co-op Power*, March 1945, 7.
4. Bud McAnally, "With Recipes on Her Mind, Blind Homemaker Enjoys Electric Range after Cooking 'Forever' on a Wood Stove," *Texas Co-op Power*, March 1967, 10.
5. "Ice Man in Maud—'I Was Pretty Blue,'" *Texas Co-op Power*, August 1945, 3.
6. "Program Explains Electricity's Help in Dairy Farming," *Texas Cooperative Electric Power*, September 1944, 4. Koon worked with registered Jersey dairy cows his whole long life. He died in Sulphur Springs in 2016. He was eighty-eight. Faith, family, and farm were his bywords, family members said.
7. "Cass 4-H Boy Uses Electrification as Demonstration Project," *Texas Cooperative Electric Power*, July 1944, 2.
8. Childs, *Farmer Takes a Hand*, 81.
9. Jack E. Nettles, "The Story of a Co-op: I—Floyd County Co-Op," *Texas Co-op Power*, January 1951, 4.
10. Milton Collier Phillips, Texas Collection, 2013, Baylor University Institute for Oral History.
11. Charles Thatcher, "Co-op Power Users Tell of Value of Electricity to Them," *Texas Co-op Power*, February 1946, 2.
12. James L. Haley, *Passionate Nation: The Epic History of Texas* (New York: Free Press, 2006), 486.
13. Bill Thompson, "Helen Hicks Recalls Changes," *Gilmer Mirror*, Special Upshur Rural Section, November 4, 1987, 15.
14. G. W. Haggard, *Texas Cooperative Electric Power*, July 1944, 2; Stephen Harrigan, *Big Wonderful Thing: A History of Texas* (Austin: University of Texas Press, 2019), 553.
15. Caro, *Path to Power*, 528.

Chapter 8

1. Joe Holley, "They All Came Home," *Houston Chronicle*, May 26, 2013.
2. Joe Holley, "Tiny Czech Community Honors Its Fallen," *Houston Chronicle*, November 7, 2014.
3. Joe Holley, "In Umbarger, Memories of POWs and Their Art Remain," *Houston Chronicle*, March 22, 2014.
4. Steve Horrell, *75 Years at Deaf Smith Electric Cooperative* (Virginia Beach, VA: Donning, 2013), 44.

5. Ben Procter, revised by Joseph G. Dawson III, "World War II," *Handbook of Texas Online*, Texas State Historical Association, originally published 1952, revised April 1, 2021, https://www.tshaonline.org/handbook/entries/world-war-ii
6. Pence, *Next Greatest Thing*, 167–68.
7. "Material Shortage Continues to Slow Co-op Building," *Texas Co-op Power*, May 1946, 1.
8. Childs, *Farmer Takes a Hand*, 107.
9. "Copper Hidden in Texas Cited as Proof REA Hinders Defense," *Dallas Morning News*, December 1, 1941.
10. *Congressional Record*, December 1, 1941, 9499–502.
11. *Dallas Morning News*, April 10, 1946.
12. "Dallas News Assails Co-ops in Special Series," *Texas Co-op Power*, May 1946, 3.
13. Childs, *Farmer Takes a Hand*, 93.
14. Studs Terkel, *Hard Times: An Oral History of the Great Depression* (New York: Pantheon Books, 1970), 230.
15. *Texas Cooperative Electric Power*, August 1944, 5.
16. "'Full Speed Ahead' Is Co-op Watchword," *Texas Co-op Power*, September 1945, 1.
17. "All-Out Production for Peace," *Texas Co-op Power*, September 1945, 2.
18. Ibid.

Chapter 9

1. Elizabeth Carpenter, "George W. Haggard: Rural Electrification Is His Religion," *Austin American-Statesman*, April 10, 1949, 1.
2. Ibid.
3. Charles M. Curfman, "The Baby Is Born," *Texas Cooperative Electric Power*, July 1, 1944, 1.
4. "A Record of Accomplishment," *Texas Cooperative Electric Power*, July 1, 1944, 2.
5. George W. Haggard, "To Hell with the Law!," *Texas Cooperative Electric Power*, November 1944, 3.
6. Bernard Rapoport, *Being Rapoport: Capitalist with a Conscience*, as told to Don Carleton (Austin: University of Texas Press, 2002), 340.
7. Dugger, *Politician*, 208.
8. Robert J. Robertson, "Montgomery, Robert Hargrove (1893–1978)," *Handbook of Texas Online*, Texas State Historical Association, February 6, 2009, https://www.tshaonline.org/handbook/entries/montgomery-robert-hargrove
9. "Give the Public Light," *Texas Cooperative Electric Power*, August 1944, 2.
10. "Co-ops under Attack; Farmers Gird for Fight," *Texas Cooperative Electric Power*, October 1944, 1.
11. "Sample of NTEA Propaganda," *Texas Cooperative Electric Power*, October 1944, 1.

12. "Co-ops Are the Essence of Free Enterprise," *Texas Cooperative Electric Power*, October 1944, 2.

13. "Ellis Sends Greetings," *Texas Cooperative Electric Power*, December 1944, 1.

14. "House Committee Hears Co-ops Attacked and Defended," *Texas Cooperative Electric Power*, December 1947, 1.

15. G. W. Haggard, "Highlights along Rural Hi-Lines," *Texas Cooperative Electric Power*, November 1944, 3.

16. "Haggard Leaves State Organization, Becomes REA Assistant Administrator," *Texas Co-op Power*, February 1948, 1.

17. "REA Deputy Chief, George W. Haggard, Talked for Congress," *Abilene Reporter-News*, May 19, 1951, 6-B.

18. "50 on Airliner Die in Colorado Plane Crash on Mountain Slope; Passengers and Crew Members on Wrecked Plane," *New York Times*, July 1, 1951, A1.

19. "Tribute Paid to Haggard," *Abilene Reporter-News*, July 2, 1951, 7-B.

20. "Rural Texas," Bill Lewis, *Texas Co-op Power*, July 1954, 1.

21. Karen Nejtek, interview by the author, December 21, 2021, Austin, TX.

22. Charles Lohrmann, "A Commitment to Quality of Life," *Texas Co-op Power*, August 2019, 8.

23. Jessica Ridge, "The Domestic Electric," *Texas Co-op Power*, August 2019, 9.

24. Kaye Northcott, "The Digital Age," *Texas Co-op Power*, August 2019, 10.

25. Joe Holley, "Exotic Energy," *Texas Co-op Power*, August 2019, 11.

26. Carol Moczygemba, "Smart Life," *Texas Co-op Power*, August 2019, 13.

27. Karen Nejtek, interview.

28. Bill Lewis, "'30' for the Editor," *Texas Co-op Power*, January 1993, 2.

Chapter 10

1. "400 Attend Post-War Conference," *Texas Cooperative Electric Power*, December 1944, 1.

2. "Major Issues Face Legislature Convening January 9, 1945," *Texas Cooperative Electric Power*, January 1945, 1.

3. "Widespread 'Spite Lines' Are Reported," *Texas Co-op Power*, July 1945, 1.

4. "Preventing 'Spite Lines,'" *Texas Co-op Power*, August 1945, 2.

5. Joe Holley, "Pappy Led the Way for Patrick," *Houston Chronicle*, October 24, 2014.

6. T. R. Fehrenbach, *Lone Star: A History of Texas and the Texans* (New York: Collier Books, 1968), 656.

7. Holley, "Pappy Led the Way."

8. Ibid.

9. Jerry Sadler, *Politics, Fat-Cats and Honey-Money Boys: The Mem-Wars of Jerry Sadler* (Santa Monica, CA: Roundtable, 1984).

10. Holley, "Pappy Led the Way."

11. Bill Minutaglio, *A Single Star and Bloody Knuckles* (Austin: University of Texas Press, 2021), 154.

12. "O'Daniel Fights REA Program," *Texas Co-op Power*, October 1947, 2.

13. "Truman Bids Nation Aid Stricken City," *Salt Lake Tribune*, April 18, 1947.

14. *Knoxville News Sentinel*, February 22, 1947.

15. Margaret Mayer, "School Heads Doubt Their Own Success," Your Capital City, *Austin American-Statesman*, September 11, 1947, 1.

16. "O'Daniel Fights REA Program," *Texas Co-op Power*, October 1947, 2.

17. "Co-ops Forge Ahead Despite Opposition from Power Companies," *Texas Co-op Power*, April 1946, 1.

18. "La Follette, Poage and Ellis Describe Power Company Fight to Cripple REA," *Texas Co-op Power*, April 1946, 4.

19. Childs, *Farmer Takes a Hand*, 103–4.

20. Ted Case, *Power Plays: The U.S. Presidency, Electric Cooperatives, and the Transformation of Rural America* (self-pub., 2013), 28.

21. Ibid., 31.

22. Ibid., 32.

23. Ibid., 34.

24. Ibid., 36.

25. "One Congressman Would Double REA Interest Rate," *Texas Co-op Power*, July 1953, 10.

26. Case, *Power Plays*, 44.

27. Ibid., 49.

28. Ibid.

29. "Lawsuit against Upshur EC," *Texas Co-op Power*, February 1955, 1.

30. Ibid.

31. *Texas Co-op Power*, April 1955, 1.

32. "Upshur EC Lawsuit," *Texas Co-op Power*, June 1955, 5.

33. "Co-ops Are Here to Stay," *Texas Co-op Power*, May 1956, 4.

34. "Texas Supreme Court Rules against Co-ops in Upshur Case—Must Resort to Legislature," *Texas Co-op Power*, March 1957, 4.

35. "Ruin for Rural Electrification," *Texas Co-op Power*, April 1957, 2.

36. Richard C. Biever, "Willie Wiredhand Turns 50," *Penn Lines*, May 2001.

37. Ibid.

38. "Reddy Kilowatt v. Willie Wiredhand," *Texas Co-op Power*, March 1956, 2.

39. "Willie Wiredhand Wins," *Texas Co-op Power*, August 1956, 8.

40. "Willie Wiredhand," *Texas Co-op Power*, February 1957, 2.

Chapter 11

1. Walter Prescott Webb, *The Great Plains* (Lincoln: University of Nebraska Press, 1931), 17, 160.

2. Elmer Kelton, *The Time It Never Rained* (New York: Doubleday, 1973; Fort Worth: Texas Christian University Press, 1984), x. Citation refers to the 1984 edition.

3. Greg Curtis, "West Is West," *Texas Monthly*, June 1999.

4. Webb, *Great Plains*, 184, 205.

5. Fred Gipson, “Eldoradoan Had Better Not Find Out Who Killed His Dogs,” *San Angelo Evening Standard*, October 30, 1939, 3.

6. Childs, *Farmer Takes a Hand*, 187.

7. “Listen to Fred Gipson Tell a Tale,” *Paris News*, January 24, 1938, 2.

8. Ibid.

9. Eddie Albin, “History of Southwest Texas Electric Cooperative, Inc.” (unpublished report, 2008), 1–2.

10. Ibid.

11. Eddie Albin, interview by the author, December 17, 2022, Eldorado, TX.

12. Albin, “History of Southwest Texas,” 2.

13. Childs, *Farmer Takes a Hand*, 187.

14. Ibid.

15. Ibid., 188.

16. Albin, “History of Southwest Texas,” 3.

17. Ibid.

18. Ibid., 12.

19. William “Buff” Whitten, interview by the author, December 17, 2022, Eldorado, TX.

20. “South Plains Electric: Progress on the March,” *Texas Co-op Power*, January 1957, 10.

21. Ibid.

22. Ibid.

23. “Remembering When,” South Plains Electric Cooperative, accessed November 23, 2022, www.spec.coop.

24. “South Plains Electric.”

25. Ibid., 10.

26. Ibid.

27. Joe Holley, “Danger Lurks in Salt Flats, as It Did a Century Ago,” *Houston Chronicle*, August 26, 2016.

28. Paul Cool, *Salt Warriors: Insurgency on the Rio Grande* (College Station: Texas A&M University Press, 2008), 59.

29. Derrill Holly, “Powering a Ranch the Size of Rhode Island,” *RE Magazine*, January 1, 2017.

30. Ibid.

31. “From the Frontier Stage,” *Texas Co-op Power*, July 1955, 3.

32. “Co-op Power and the Last Frontier,” *Texas Co-op Power*, January 1954, 1.

33. Ibid.

34. *The WPA Guide to Texas* (Austin: Texas Monthly Press, 1986), 619.

35. Christian Wallace, “The Last Stand at Alamo Village,” *Texas Monthly*, March 2018.

36. Ibid.

37. “Alamo in Brackettville,” *Texas Co-op Power*, June 1958, 4.

38. Wallace, “Last Stand.”

Chapter 12

1. Joe Holley, "What Hath Philo Wrought?," *Texas Co-op Power*, May 1999, 12.
2. "The History of Television," https://www.thehistoryoftv.com.
3. "The Farnsworth Chronicles," https://farnovision.com/chronicles/.
4. Michael Pollak, "Screen Grabs; Inside the Soap Opera of Television's Early Days," *New York Times*, January 18, 2001, G8.
5. "¾ of Farms Have Radios, but Bathrooms Still Scarce," *Texas Co-op Power*, December 1949, 4.
6. Richard Schroeder, *Texas Signs On: The Early Days of Radio and Television* (College Station: Texas A&M University Press, 1998), 86.
7. Ibid., 87.
8. "TV-Eye View of a Texas Farm Family," *Texas Co-op Power*, July 1950, 1.
9. "New TV Stations in Texas Await FCC Decision on 'Standard System,'" *Texas Co-op Power*, July 1950, 4.
10. "Co-op Members Think TV 'Just the Thing' for Farm," *Texas Co-op Power*, July 1950, 4.
11. "New TV Stations in Texas."
12. "Television in Texas," *Texas Co-op Power*, December 1953, 6.
13. Carol Moczygemba, "Who's in Charge Here?," Texas *Co-op Power*, May 1999, 8.
14. "Television in Texas."
15. Moczygemba, "Who's in Charge Here?"

Chapter 13

1. Joe Holley, "Marlin's Restorative Waters Still Flow, but Glory Is Gone," *Houston Chronicle*, May 14, 2013.
2. Joe Holley, "In Paducah, 'the Only Thing Growing Is the Cemetery,'" *Houston Chronicle*, October 20, 2013.
3. Rachel Richards, "Looking Back at My Little Town," *Texas Co-op Power*, September 1999, 14.
4. Carol Moczygemba, "Polishing Up the 'Gem City of the Blackland Belt,'" *Texas Co-op Power*, October 1998.
5. Ronald R. Kline, *Consumers in the Country: Technology and Social Change in Rural America* (Baltimore: Johns Hopkins University Press, 2000), 216.
6. "Why Do They Leave the Farm?," *Texas Co-op Power*, March 1948, 2.
7. "Farm Population Declining," *Texas Co-op Power*, January 1952, 2.
8. "Finally the Rains Came," *Texas Co-op Power*, June 1957, 10.
9. Michael E. Sloan, "Propaganda, Prejudice and the Rural Exodus," *Texas Co-op Power*, June 1961, 3.
10. Richard M. Hausler, "Rural America's Great Loss . . . Its Young People," *Texas Co-op Power*, November 1961, 15.
11. Robert D. Partridge, "Rural/Urban Crisis," *Texas Co-op Power*, May 1968, 11.

12. Matthew Sanderson, "Ag-rich Counties Still Losing Population," *Rural Messenger*, September 14, 2021.

13. Sarah Self-Walbrick, "The 'Big Empty' Could Dilute Rural Texans' Representation, Experts Say," Houston Public Media, September 29, 2021.

14. Noah Wicks, "Redistricting Could Reshuffle House Ag Committee Membership," *Agri-Pulse*, January 19, 2022.

15. Robert Cushing, "Tracing the Divide: Urban and Rural Voting Preferences Started to Diverge in the 1970s," *Daily Yonder*, May 20, 2021.

16. Sara June Jo-Saebo, "Commentary: Beyond Belief—the Comforts of a Country Church," *Daily Yonder*, February 3, 2022.

17. Joe Holley, "There's No Bowling Alone in Blanco," *Houston Chronicle*, September 23, 2018.

18. Bob Buckel, communications and media representative, Tri-County EC, email to author, May 18, 2022.

19. Ibid.

20. Mike Williams, interview by the author, May 21, 2021, Austin, TX.

Chapter 14

1. Dick Wilson, "The Attack on G&T Electric Loans," *Texas Co-op Power*, September 1962, 2.

2. "Power Trust Seeking to Sabotage Co-op Programs by Cutting Off Its Source of Wholesale Power," *Texas Co-op Power*, December 1945, 1.

3. Ibid.

4. Wilson, "Attack on G&T Electric Loans."

5. G. W. Haggard, "Along Rural Hi-Lines," *Texas Co-op Power*, November 1947, 1.

6. Quoted in *Brazos Electric Power Cooperative Annual Report*, 1986, Brazos Electric Cooperative, www.brazoselectric.com

7. Ibid.

8. *Rayburn Country Electric Cooperative: 40 Years of Maximizing Member Value* (Brookfield, MO: Dunning, 2019), 16.

9. Ibid., 19.

10. Ibid., 20.

11. Ibid., 30.

12. Ibid., 28.

Chapter 15

1. "A Phone Executive Assails Bell System, in His Suicide Note," *New York Times*, November 19, 1974, 30.

2. Tom Curtis, "Jury Awards $3 Million to Bell Plaintiffs," *Washington Post*, September 13, 1977.

3. "49 Representatives Backing Utility Commission Measure," *San Angelo Standard-Times*, January 31, 1975, 1.

4. Associated Press, January 30, 1975.

5. "Texas Legislature Creates Public Utility Commission in an 11th-Hour Decision," *Texas Co-op Power*, July 1975, 6.

6. Walter Richter, "Poor Richard's Almanac," *Texas Co-op Power*, July 1975, 6.

7. Randall S. Mallory, "Regulation: What the New Texas Public Utility Law Promises for You and Your Cooperative," *Texas Co-op Power*, August 1975, 6.

8. Richter, "Poor Richard's Almanac," 6.

9. "Service Area Certification Top Topic for State REC Meeting," *Texas Co-op Power*, August 1975, 2.

10. Johnson, *And There Was Light*, 130.

11. Ibid., 131.

Chapter 16

1. Randall S. Mallory, "Fred McClure, 'A Traveling FFA Salesman,'" *Texas Co-op Power*, July 1973, 2.

2. Mike Williams, interview by the author, February 15, 2022, Austin, TX.

3. Kate Aranoff, "Bringing Power to the People: The Unlikely Case for Utility Populism," *Dissent*, Summer 2017.

4. Wade Rathke, "Speak Your Piece: Bring Power Back to the People," *Daily Yonder*, May 19, 2016.

5. Shirley Jackson, phone interview by the author, October 14, 2022, Gilmer, TX.

6. Richard West, "The Petrified Forest," *Texas Monthly*, April 1978.

7. Richard McClure, phone interview by the author, April 14, 2022, College Station, TX.

8. Kurt Johnson, "Weather Analyst Discovers Electricity Played a Key Role in Central Texas Blackout," *Texas Co-op Power*, August 1983, 2.

9. "Round Table Speaker Views 'Man's World,'" *Texas Co-op Power*, November 1970, 10.

10. "A Friend in Need: Mrs. J. P. Lorenz, Sr.," *Texas Co-op Power*, November 1960, 16.

11. Kline, *Consumers in the Country*, 216.

12. "Swisher Co-op: Story of Struggle and Success," *Texas Co-op Power*, September 1956, 8.

13. Ibid.

14. "Sarah Sears: Just the Right Person for the Job," *Texas Co-op Power*, February 1994, 12.

15. Ibid.

16. Moreland, *Dreams Come True*, 194.

17. Aranoff, "Bringing Power to the People."

18. Ibid.

19. Dan Charles, "North Carolina Electric Cooperative Aims to Make New Technologies Accessible to All," *Weekend Edition Sunday*, National Public Radio, March 21, 2021.

20. Victoria A. Rocha, "Diversity, Equity, Inclusion: So Our Entire Community Can Flourish," *RE Magazine*, January 1, 2021, https://www.cooperative.com/remagazine

21. Ibid.

Chapter 17

1. Garry Wills, *Reagan's America: Innocents at Home* (New York, Doubleday, 1983), 11.
2. Ronnie Dugger, *On Reagan: The Man and His Presidency* (New York: McGraw Hill, 1983), 5.
3. Ibid., 14.
4. Ibid., 15.
5. Case, *Power Plays*, 103.
6. Ibid., 108.
7. Norman M. Clapp, interview by T. H. Baker, December 6, 1968, 15, Oral History Transcript, LBJ Library Oral Histories.
8. Seth S. King, "They Lighted Up the Family Farm," *New York Times*, March 29, 1981, sec. 3, p. 4.
9. Dick Pence, "Rural Electrics Face Financing Crisis," *Texas Co-op Power*, February 1973, 3.
10. Graham Howe, "Rally Urges Return of Direct Loans," *Texas Co-op Power*, March 1973, 2.
11. Robert Partridge, "Bad News for Rural Americans," *Texas Co-op Power*, March 1981, 2.
12. William Greider, "The Education of David Stockman," *Atlantic*, December 1981.
13. Case, *Power Plays*, 11.
14. Partridge, "Bad News for Rural Americans," *Texas Co-op Power*, March 1981, 2.
15. "Shameful Effort," *Texas Co-op Power*, September 1982, 2.
16. Case, *Power Plays*, 115.
17. Ibid., 119.
18. Ibid., 131.
19. *Texas Co-op Power*, December 1992.
20. Jim Morriss, "There's a Funny Taste in the Deficit Stew," *Texas Co-op Power*, April 1993, 2.
21. Case, *Power Plays*, 139.
22. Ibid., 141.
23. Ibid., 142.

Chapter 18

1. Peg Champion, "Electrifying Lawmakers," *Texas Co-op Power*, March 1999, 3.
2. Joe Holley, "Let the Juice Flow," *Texas Co-op Power*, March 1999, 6.

3. Champion, "Electrifying Lawmakers."
4. Mike Williams, "With Sibley, It's Been a Good Trip So Far," *Texas Co-op Power*, March 1999, 5.
5. Eric Craven, interview by the author, October 15, 2022, Austin, TX.
6. Williams, "With Sibley."
7. Holley, "Let the Juice Flow."
8. Ibid.
9. Elizabeth Souder, "Pat Wood Helped Create Texas' Electricity Market, Then He Lost Power for 36 Hours in the Freeze," *Dallas Morning News*, July 18, 2021.
10. Mike Williams, interview by the author, February 7, 2022, Austin, TX.
11. Ibid.
12. Ibid.
13. Caro, Path to Power, 517.
14. Holley, "Let the Juice Flow."
15. Ibid.
16. Williams, interview.
17. Ibid.
18. Holley, "Let the Juice Flow."
19. Andrew Weber, "Here's How Texas Lawmakers (and Enron) shaped the State's Electrical Market," KUT 90.5, July 23, 2021.
20. Holley, "Let the Juice Flow."
21. Weber, "Here's How Texas Lawmakers."
22. Ibid.
23. Holley, "Let the Juice Flow."
24. Williams, interview.
25. Holley, "Let the Juice Flow."
26. Weber, "Here's How Texas Lawmakers."
27. Ibid.
28. Holley, "Let the Juice Flow."
29. Weber, "Here's How Texas Lawmakers."
30. Holley, "Let the Juice Flow."
31. Mike Williams, "A Consumer Bill of Rights," *Texas Co-op Power*, April 1999, 5.
32. Weber, "Here's How Texas Lawmakers."
33. Ibid.
34. Ibid.
35. Mike Williams, "Diving into Deregulation," *Texas Co-op Power*, July 1999, 5.
36. Mike Williams, "Be Prepared," *Texas Co-op Power*, November 1999, 5.
37. Ibid.

Chapter 19

1. Leo Janos, "The Last Days of the President," *Atlantic*, July 1973.
2. Norman M. Clapp, interview by T. H. Baker, December 6, 1968, 15, Oral History Transcript, LBJ Library Oral Histories.

3. Robert Caro, *Master of the Senate*, vol. 3 of *The Years of Lyndon Johnson* (New York: Alfred A. Knopf, 2002), 922.

4. Dugger, *Politician*, 150.

5. Navigant Consulting, *Pedernales Electric Cooperative, Inc., Report of Investigation*, December 15, 2008, 290, 30.

6. "Former Hays County Judge William 'Bud' Burnett Dies at 82," *San Marcos Daily Record*, December 31, 2020.

7. Navigant Consulting, *Pedernales Electric Cooperative*, 250.

8. Ibid., 79.

9. Ibid., 66.

10. Eric Craven, interview by the author, October 15, 2022, Austin, TX.

11. Jim Morriss, "Did You Vote?," *Texas Co-op Power*, October 1993, 2.

12. Claudia Grisales, "Election Procedures under Review, Utility Officials Say," *Austin American-Statesman*, August 4, 2007.

13. Claudia Grisales, "Missing Are Dozens of Large Charges on Pedernales Credit Cards," *Austin American-Statesman*, January 6, 2008.

14. State senator Troy Fraser, "Testimony Before the House Committee on Oversight and Government Reform," June 26, 2008.

15. Navigant Consulting, *Pedernales Electric Cooperative*, 21–22.

16. Ibid., 60.

17. Jodi Lehman, "Senate Begins Inquiry into Unregulated Cooperatives," *Horseshoe Bay Beacon*, April 3, 2008.

18. "Findings Could Speed Up Grand Jury Investigation," *Austin American-Statesman*, December 17, 2008.

19. Navigant Consulting, *Pedernales Electric Cooperative*, 152–95.

20. Ibid., 23.

21. "Generate Some Sunshine Power," *Austin American-Statesman*, March 17, 2011.

22. Fraser, "Testimony Before the House Committee."

23. "Utility Severing Former Business Ties with Family," *Austin American-Statesman*, August 17, 2008.

24. Patrick Cox, interview by the author, July 12, 2022, Driftwood, TX.

25. "Outgoing Chief: Pedernales on Solid Footing," *Austin American-Statesman*, June 22, 2010.

26. "Public Citizen Applauds Leadership, Vision of New PEC Board," *Public Citizen*, June 30, 2009.

27. Navigant Consulting, *Pedernales Electric Cooperative*, 37.

28. Roy Reed, "President Urges Firmness on War; In First Visit to Dallas Since Assassination, He Sees a Turning Point in Vietnam," *New York Times*, February 28, 1968, 1.

29. Case, *Power Plays*, 74.

30. Reed, "President Urges Firmness."

31. Case, *Power Plays*, 74–75.

Chapter 20

1. J. Marvin Hunter, *Pioneer History of Bandera County—Seventy-Five Years of Intrepid History* (Bandera, TX: Hunter's Printing House, 1922), 26.
2. Ibid., 40.
3. Jeff Siegel, "Broadband across Texas," *Texas Co-op Power*, August 2017.
4. Paul Flahive, "'You're Doing This to Allow Your Communities to Survive': Rural Electric Co-ops on Broadband," Texas Public Radio, May 17, 2019.
5. TEC report, March 15, 2021. [These reports are internal documents that go only to TEC employees and board members.]
6. TEC report, July 1, 2021.
7. TEC report, April 1, 2021.
8. TEC report, February 1, 2020.
9. TEC report, March 1, 2021.
10. TEC report, November 16, 2020.
11. TEC report, December 23, 2020.

Chapter 21

1. Norman Merchant, "Power Failure: How a Winter Storm Pushed Texas into Crisis," Associated Press, February 21, 2021.
2. Kara Norton, *Nova*, PBS, March 25, 2021.
3. Charles Blanchard, "The Great Texas Blackout Was Caused by a Failure to Ensure Supplies of Natural Gas, *Texas Monthly*, March 3, 2021.
4. Synda Smith, phone interview by the author, February 25, 2022, Coleman, TX.
5. Bill Hancock, "Lessons Learned from 2021 Winter Storm Uri," *Runnels County Register*, January 10, 2022.
6. Smith, phone interview.
7. *Resilience Is Reliance*, Golden Spread Electric Cooperative, Inc., 2021 annual report, 6.
8. Blanchard, "Great Texas Blackout."
9. "Brazos Electric Power Cooperative, Inc. Files for Chapter 11 Financial Restructuring," Brazos Electric Cooperative newsletter, March 1, 2021.
10. Llewellyn King, "Uri Fallout: Texas' Rayburn Electric Is First Co-op to Securitize Uri Debt; Still Expected to Sue," *Forbes*, February 15, 2022.
11. Mauri Montgomery, "United Storm Response Deemed Best Practice," United Cooperative Services newsletter, April 1, 2021.
12. Mauri Montgomery, "United Issues Statement Regarding Brazos Electric Bankruptcy Filing," United Cooperative Services newsletter, March 1, 2021.
13. Norton, *Nova*.
14. Chris Burrows, "That's What Co-ops Are For," *Texas Co-op Power*, June 2021.

Chapter 22

1. *The Electric Cooperative Purpose: A Compass for the 21st Century*, report of the NRECA 21st Century Cooperative Committee, 2013, 8.
2. Camille Wheeler, "Don't Do Mild," *Texas Co-op Power*, March 2012, 8.
3. Mike Williams, "Chairman's Letter," *Electric Cooperative Purpose: A Compass for the 21st Century*, 2013, v.
4. "Executive Summary," *Electric Cooperative Purpose*, xii.
5. Ibid.
6. Ibid.
7. "Case Studies," *Electric Cooperative Purpose*, 23.
8. Ibid., 43.
9. Ibid., 57.
10. Ibid., 60.
11. Wheeler, "Don't Do Mild."
12. Llewellyn King, "Texas Power Firm Hit with 2.1 Billion, Texas Co-ops Still Worried about the Next Big Freeze," *Forbes*, January 6, 2022.
13. Andrew Freedman, "Deadly Texas Blackout Shows Our Vulnerability to Coming Climate Extremes," *Washington Post*, February 22, 2021.
14. Jeremy Deaton, "Business Booms at Climate Risk Start-Up as Threat from Extreme Weather Grows," *Washington Post*, May 27, 2021.
15. Kasha Patel, "Electricity Blackouts Could Be Avoided across the Nation by Switching to Solar, Wind and Water Energy Resources, Report Says," *Washington Post*, March 3, 2022.
16. Ibid.
17. Reed Karaim, "Technology Trends: 'Co-ops 25,' How Rapid Change in the Industry Today Will Play Out over the Next Quarter Century," *RE Magazine*, December 1, 2020.
18. "Rural Electric Co-ops Face Barriers to Going Renewable," Yale Climate Connections, April 26, 2022.
19. Mike Williams, interview by the author, January 23, 2022, Austin, TX.

BIBLIOGRAPHY

Arnst, Rita, and Julia Bedsole. *Nueces Electric Cooperative: Celebrating 75 Years.* Virginia Beach, VA: Donning, 2014.

Caro, Robert. *The Path to Power.* Vol. 1 of *The Years of Lyndon Johnson.* New York: Alfred A. Knopf, 1983.

———. *Master of the Senate.* Vol. 3 of *The Years of Lyndon Johnson.* New York: Alfred A. Knopf, 2002.

Case, Ted. *Poles, Wires and War: The Remarkable Untold Story of Rural Electrification and the Vietnam War.* Self-published, 2017.

———. *Power Plays: The U.S. Presidency, Electric Cooperatives, and the Transformation of Rural America.* Self-published, 2013.

Childs, Marquis. *The Farmer Takes a Hand: The Electric Power Revolution in Rural America.* Garden City, NY: Doubleday, 1952.

Dugger, Ronnie. *On Reagan: The Man and His Presidency.* New York: McGraw-Hill, 1983.

———. *The Politician: The Life and Times of Lyndon Johnson; The Drive for Power, from the Frontier to Master of the Senate.* New York: W. W. Norton, 1982.

Freeberg, Ernest. *The Age of Edison: Electric Light and the Invention of Modern America.* New York: Penguin Press, 2013.

Goodwyn, Lawrence. *Democratic Promise: The Populist Moment in America.* New York: Oxford University Press, 1976.

Haley, James L. *Passionate Nation: The Epic History of Texas.* New York: Free Press, 2006.

Hardeman, D. B., and Donald C. Bacon. *Rayburn: A Biography.* Austin: Texas Monthly Press, 1987.

Harrigan, Stephen. *Big, Wonderful Thing: A History of Texas.* Austin: University of Texas Press, 2019.

Hollandsworth, Skip. *The Midnight Assassin: Panic, Scandal, and the Hunt for America's First Serial Killer.* New York: Henry Holt, 2015.

Horrell, Steve. *75 Years at Deaf Smith Electric Cooperative.* Virginia Beach, VA: Donning, 2013.

Houston Advanced Research Center and the Institute for Energy, Law and Enterprise, University of Houston. *Guide to Electric Power in Texas.* 3rd ed. Houston: Houston Advanced Research Center and the Institute for Energy, Law and Enterprise, University of Houston, 2003.

Hunter, J. Marvin. *Pioneer History of Bandera County: Seventy-Five Years of Intrepid History.* Bandera, TX: Hunter's Printing House, 1922.

Johnson, Neal. *And There Was Light: The History of Kaufman County Electric Cooperative, 1937–1997.* In collaboration with Ray Raymond. Austin: Texas Electric Cooperatives, 1998.

Kelton, Elmer. *The Time It Never Rained.* New York: Doubleday, 1973; Fort Worth: Texas Christian University Press, 1984.

Kline, Ronald R. *Consumers in the Country: Technology and Social Change in Rural America.* Baltimore: Johns Hopkins University Press, 2000.

Lind, Michael. *Land of Promise: An Economic History of the United States.* New York: Harper, 2012.

Marsh, Brittany, and Laurie Putska. *Victoria Electric Cooperative, Inc.: Celebrating 75 Years.* Virginia Beach, VA: Donning, 2013.

Moreland, Brian K. *Dreams Come True: The 75 Year History of Hilco Electric Cooperative, Inc., 1937–2012.* Itasca, TX: Hilco Electric Cooperative, 2010.

Morris, Edmund. *Edison.* New York: Random House, 2019.

Pence, Richard, ed. *The Next Greatest Thing: 50 Years of Rural Electrification in America.* Washington, DC: National Rural Electric Cooperative Association, 1984.

"Pioneer History of Bandera County: Seventy-Five Years of Intrepid History." The Portal to Texas History, University of North Texas Libraries. Accessed June 5, 2022. https://texashistory.unt.edu/ark:/67531/metapth27720/

Railey, James H., project director. *Rayburn County Electric Cooperative: 40 Years of Maximizing Member Value.* Virginia Beach, VA: Donning, 2019.

Rapoport, Bernard. *Being Rapoport: Capitalist with a Conscience.* As told to Don Carleton. Austin: University of Texas Press, 2002.

Reid, Debra A. *Reaping the Greater Harvest.* College Station: Texas A&M University Press, 2007.

Schroeder, Richard. *Texas Signs On: The Early Days of Radio and Television.* College Station: Texas A&M University Press, 1998.

Terkel, Studs. *Hard Times: An Oral History of the Great Depression.* New York: Pantheon Books, 1970.

Thompson, David J. *Weavers of Dreams: Founders of the Modern Co-operative Movement.* Davis: Regents of the University of California, 1994.

Webb, Walter Prescott. *The Great Plains.* Lincoln: University of Nebraska Press, 1931.

Wills, Garry. *Reagan's America: Innocents at Home.* New York: Doubleday, 1987.

INDEX

ABOUT THE AUTHOR

Joe Holley has been the "Native Texan" columnist for the *Houston Chronicle* since 2013. A native Texan himself—from Waco—he's been an editorial-page editor in San Diego, California, a contributor to *Texas Monthly*, a speechwriter for Governor Ann Richards, editor of *Texas Co-op Power*, a staff writer for the *Washington Post*, and an editorial writer for the *Chronicle* from 2012 to 2017. He was a Pulitzer Prize finalist in 2017 for a series of editorials on gun control and the Texas gun culture, a Pulitzer Prize finalist in 2023 for editorials on the mass shooting at an elementary school in Uvalde, Texas, and a Pulitzer Prize winner in 2022, as part of the *Houston Chronicle* editorial team that produced a series of editorials on Donald Trump's "Big Lie."

An alumnus of Abilene Christian University, UT-Austin, and the Columbia University Graduate School of Journalism, he's the author of six books, including *Hometown Texas*, a collection of his weekly "Native Texan" columns; *Hurricane Season: The Unforgettable Story of the Houston Astros and the Resilience of a City*; and *Sutherland Springs: God, Guns, and a Small Texas Town*, published in 2020 and recipient of the 2021 Carr P. Collins Award, presented by the Texas Institute of Letters in recognition of the year's best work of nonfiction. The book explores the aftermath of the mass shooting at the Baptist church in Sutherland Springs, Texas, on November 5, 2017.